THE BURDEN OF EMPIRE

PERSPECTIVES ON IMPERIALISM AND COLONIALISM

IAN COPLAND

Melbourne

OXFORD UNIVERSITY PRESS

FOR M.N.G.

OXFORD UNIVERSITY PRESS AUSTRALIA

Oxford New York
Athens Auckland Bangkok Bombay
Calcutta Cape Town Dar es Salaam Delhi
Florence Hong Kong Istanbul Karachi
Kuala Lumpur Madras Madrid Melbourne
Mexico City Nairobi Paris Singapore
Taipei Tokyo Toronto

and associated companies in
Berlin Ibadan

OXFORD is a trade mark of Oxford University Press

First published 1990
Reprinted 1996

National Library of Australia
Cataloguing-in-Publication data:

Copland, Ian, 1943–
The burden of empire: perspectives on imperialism and colonialism.

Bibliography.
Includes index.
ISBN 0 19 553208 2.

1. Imperialism. 2. Colonies—Asia. 3. Asia—History.
1. Title.

325.32

Designed by Steve Randles
Typeset by Syarikat Seng Teik Sdn. Bhd., Malaysia
Printed by McPherson's Printing Group
Published by Oxford University Press,
253 Normanby Road, South Melbourne, Australia

Contents

Prologue

Imperialism! Hang
the word! It
buzzes on my
noddle,
Like bumble-bees in
clover time,
The talk on 't's
mostly twaddle.

***Punch*, 1877**

What is 'imperialism'? What is 'colonialism'? How, and to what extent, do they differ? To begin, let us try to determine what the words originally meant when they entered the language in the late 19th century.

Imperialism is clearly derived from the word 'empire', which comes from the Latin word *imperium*, meaning 'rule' or 'authority'. Thus, imperialism can be defined as the *imposition* (usually by force) *of the rule of one state or country over another*. More simply, imperialism can be regarded as the *activity of empire-building*. However, any definition of imperialism should also include the motivations and attitudes of the colonisers. Hence, the term is used here to describe the *acquisition, by a state or country, of power or influence over another, and the ideas and attitudes which supported and justified this activity.*

Colonialism is derived from the Greek word meaning 'settlement'. However, as European colonial empires grew more and more Asian- and African-centred, the original meaning of the term became obscured, and 'colonialism' gradually came to refer to *those parts of the non-Western world which were under direct European rule*. More recently, still, colonialism has come to be thought of as something inherently evil and oppressive: a system geared essentially to *economic exploitation*. However, it should not be assumed that European rule did nothing else but oppress and exploit. Until this 'exploitative' viewpoint is tested against the evidence, its truth cannot be assumed. Historians must hear the accused: guilt should not be presumed simply because a charge has been laid. This book, therefore, will define colonialism as *the political, social and economic penetration—for good or ill—of non-European societies.*

What is the connection between the two terms? Imperialism is about the *acquisition* of power and influence, colonialism is about *what happens* to native societies which come under European rule. Thus colonialism is a possible *outcome*—the next stage—of imperialism.

Each chapter of this book focuses on particular aspects of imperialism and colonialism. The early chapters consider the hows and whys of empire-building and the changing face of imperialism. The middle chapters illustrate colonial policies and the effects of Westernisation on the native peoples. The later chapters look at the reasons behind decolonisation with an overall assessment of imperialism and colonialism.

An important structural feature of the book is its emphasis on 'documents'. As far as possible, the documents have been allowed to speak for themselves. While questions have been provided to provoke thought and stimulate further enquiry, readers have been left to draw their own conclusions about what is important.

1 *The Roots of Expansion*

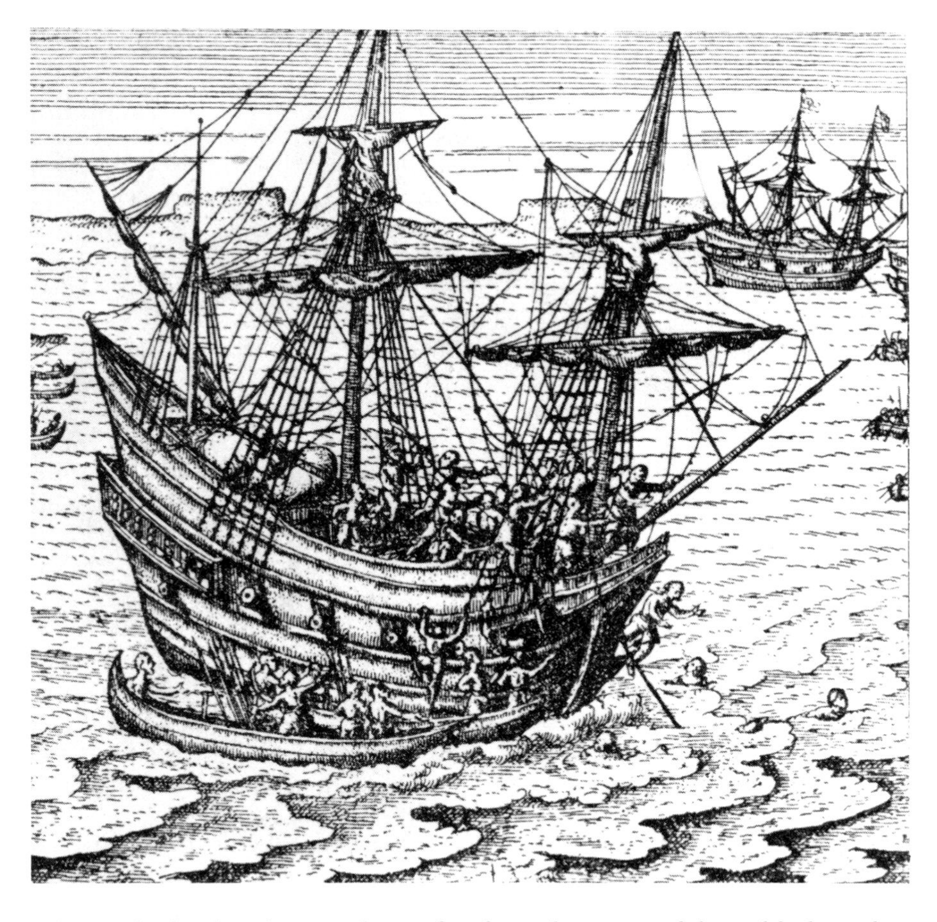

They asked what he sought so far from home, and he told them he came in search of Christians and of spices . . .

the logbook of the *San Rafael*, 1498

THIS CHAPTER FOCUSES MAINLY ON the first phase of European global expansion: the so-called Age of Discovery which lasted from the 15th century to the 18th century. In particular, it looks at the political and social aspirations which motivated this expansion. What drove European men (and later, women) to brave hardship and death in remote foreign lands? Was it for God that they went? Or for material gain? Or from a sense of duty? Or merely out of curiosity?

KEY DATES

1295 Venetian adventurer Marco Polo returns home after 24 years in China and Central Asia.

1415 Capture of the 'Moorish' (Arab) stronghold of Ceuta (Morocco) by Prince Henry ('the Navigator') of Portugal (1394–1460).

1498 Portuguese expedition led by Vasco da Gama (*c.* 1469–1524) lands at Calicut, India, opening up the sea-route to Asia.

1511 Malacca (Malaysia) captured by a Portuguese force under Alfonso d'Albuquerque (Viceroy of the Indies 1509–15), the first major Asian city to fall into European hands.

1519–21 Spanish expedition led by Ferdinand Magellan (*c.* 1480–1521) and, after his death in the Philippines, by del Cano, is the first to sail around the world.

1549 Jesuit missionary Francis Xavier (1506–52) brings Christianity to Japan.

1571 Capture of Manila begins 300 years of Spanish colonialism in the Philippines.

1601 English East India Company founded in London.

1602 Vereenigde Oost-Indische Compagnie (Dutch East India Company) founded in Amsterdam.

1641 All foreigners expelled from Japan under the *sakoku* (closed) policy introduced by Shogun Tokugawa Iemitsu (*r.* 1622–51).

1698 English East India Company granted *zamindari* (landlord) rights to the village of Kalikata (Calcutta) in Bengal.

1736 Englishman John Harrison (1693–1776) invents the first reliable ship's clock—the chronometer—making possible the accurate calculation of longitude at sea.

ISSUES AND PROBLEMS

The contemporary world of the late 20th century is a 'global village' dominated by Western ideas and inventions, so it is difficult to imagine that there was ever a time when Europe did not represent the core of human civilisation. In fact the 'European

Age' is relatively recent—at best, a feature of the last 400 years. Before that, Europe was not so much dominant as dominated. For nearly 1000 years after the fall of the Roman Empire in the 4th century AD, the northern parts of the European subcontinent were repeatedly ravaged by waves of fierce, westward-migrating peoples from the central Asian steppes—Goths, Vandals, Huns, Bulgars and Mongols. And during the 8th century, large parts of the Mediterranean world, including the whole of the Iberian Peninsula, were overrun and colonised by Islamicised Arabs.

The Islamic empire straddled the Mediterranean and constituted a formidable barrier to European expansion south and east. However, the biggest deterrent to Western expansion during the Medieval period was not the presence of the Arabs (or 'Moors'), but Christian teaching about the geography of the world. In fact, much accurate knowledge about the inhabited world was available in the books and maps of ancient Greek scholars such as Ptolemy of Alexandria (d. AD 141), but these 'pagan' writings were condemned as **heretical** by the Catholic Church. Instead, the Church proclaimed a view of the world which fitted the teachings of the Old Testament. The earth was said to be (as indeed it looked) a flat, rectangular, box-like object with finite boundaries. Moreover, the outer limits of this world, beyond the familiar locations of the Mediterranean (or 'middle') Sea, were said to be dangerous in the extreme. Somewhere to the east lay the Garden of Eden, but the route to this earthly paradise lay through trackless deserts and impenetrable jungles teeming with deadly serpents; to the south lay a zone of fire; and elsewhere the margins of the world were populated by terrifying subhumans—pygmies, dog-men, and people with four eyes.

heretical unorthodox, contrary to accepted belief, usually with respect to religion

Of course, not everyone in Europe believed these fantasies, and in the 13th century, a few hardy adventurers, taking advantage of the hospitality of the Mongol Empire, ventured overland as far as China (source 1.7). Yet sea-voyages in particular were discouraged by fear of what lay over the horizon. By the early 1400s, Genoese mariners had discovered the Canary Islands and Portuguese seafarers had reached Cape Bojador on the north-east coast of Africa. But exploration stopped there, halted by superstition. As a contemporary wrote:

> . . . this much is clear, that beyond this Cape there is no race of men nor place of inhabitants. . . . the sea is so shallow that a whole league from land it is only a fathom deep, while the currents are so terrible that no ship having once passed the Cape, will ever be able to return.[1]

Surrounded by the power of Islam and inhibited by fears of the perils of sea travel, Europeans of the Middle Ages also lacked most of the resources necessary for world-conquest. Technologically, their society was backward. Their firearms were unreliable, their ships were tiny and difficult to manoeuvre against the wind, and their medical knowledge was primitive. Maps of everything beyond the Mediterranean were sketchy or non-existent. Until the 17th century, they had no telescopes or barometers and, until the 18th century, no accurate way of calculating longitude. Economically, too, Europe was in poor shape. Between 1347 and 1377, approximately 40 per cent of the entire population was killed by plague in what became known to history as the Black Death, and mortality levels remained very high until well into the following century. In the process, trade shrank and manufactures stagnated. Not until the advent of the Industrial Revolution, around 1750, did Europe gain a decisive economic advantage over the rest of the world.

Portugal played a leading role in the process of European expansion. In the words of the famous Portuguese poet, Camoens, who chronicled the history of the epic voyages of the 15th and 16th centuries in his ballad, the *Lusiads*: 'This is a story of heroes who, leaving their native Portugal behind them, opened a way to Ceylon, and further across seas no [European] man has sailed before'.[2] By any standards this was a remarkable achievement. But it becomes even more amazing when one considers the country from whence they came. At the beginning of the 16th century, Portugal was a small, sparsely populated kingdom beset by recurrent food shortages. How did this weak, backward nation manage, in the space of little more than a century, to throw off the Islamic yoke, forge a sea-road to Asia and annex for itself a large share of the rich commerce of the Indian Ocean?

THE DEBATE

Historians have put forward four basic explanations for the sudden, global European expansion of the late 15th and 16th centuries. The first of these is couched in *strategic* terms: expansion to Asia, it is said, had its roots in 'thirty centuries of strenuous border warfare against the Moors'[3] and was undertaken with a view to making contact with potential allies such as the legendary Christian king Prester John (source 1.5). The second explanation emphasises *curiosity*: 'the lure of the unknown'.[4] The third stresses *material factors*, most importantly the pursuit of quick riches through plunder. As Italian historian Carlo Cipolla puts it, 'religion provided the pretext and gold the motive'.[5] And the fourth focuses on the global impacts of the *scientific and military inventions* of the 16th century, which together turned Europeans into the world's most efficient, and perhaps callous, killers. The first three of these explanations are examined below. The last will be discussed in chapter 3.

Dangers and Discomforts of Exploration

In the 16th and 17th centuries, long distance sea-travel was a hazardous business, with losses from shipwreck on the Cape route averaging around 50 per cent. Lack of fresh food and water, which brought on scurvy (extreme vitamin C deficiency) and other diseases, was perhaps the biggest problem that these early mariners faced; and in this respect the epic Magellan–del Cano expedition of 1519–22, crossing seas where no Europeans had gone before, fared worse than most. In source 1.1, crew member Antonia Pigafetta of Vincenza (1480/90?–1534) recalls the trials he and his shipmates faced crossing the vast Pacific.

In source 1.2, another common maritime hazard, an electrical storm, is described by Edmund Barker, the First Lieutenant of the English ship *Edward Bonaventure*, commanded by Captain (later Sir James) Lancaster. The vessel survived and went on to round the Cape, the first English ship to do so from West to East. Eventually it reached Penang. But by the time Lancaster returned home, in 1594, only 25 of his crew of 198 were still alive. Moreover, reaching Asia was only the first step. Once there, Europeans had to cope with a hostile climate, strange food, killer diseases such as malaria, typhoid and cholera, and—worst of all perhaps—isolation from kith and kin.

Yet, the American historian, Holden Furber, reminds us that conditions in the East may not have been all that much different from those which the lower classes had to put up with at home even during the 19th century (source 1.3). If so, their readiness to brave the hazards of the tropics becomes easier to comprehend.

uropean exploration nd discovery, 13th–16th enturies.

he Polo Brothers 1255–69

artolemo Diaz 1487–88

hristopher Columbus 1492

asco da Gama 1497–98

ebastian Cabot 1505

erdinand Magellan–de Cano 1519–22

rancis Xavier 1540–52

rancis Drake 1577–80

ornelis Houtman 1595–97

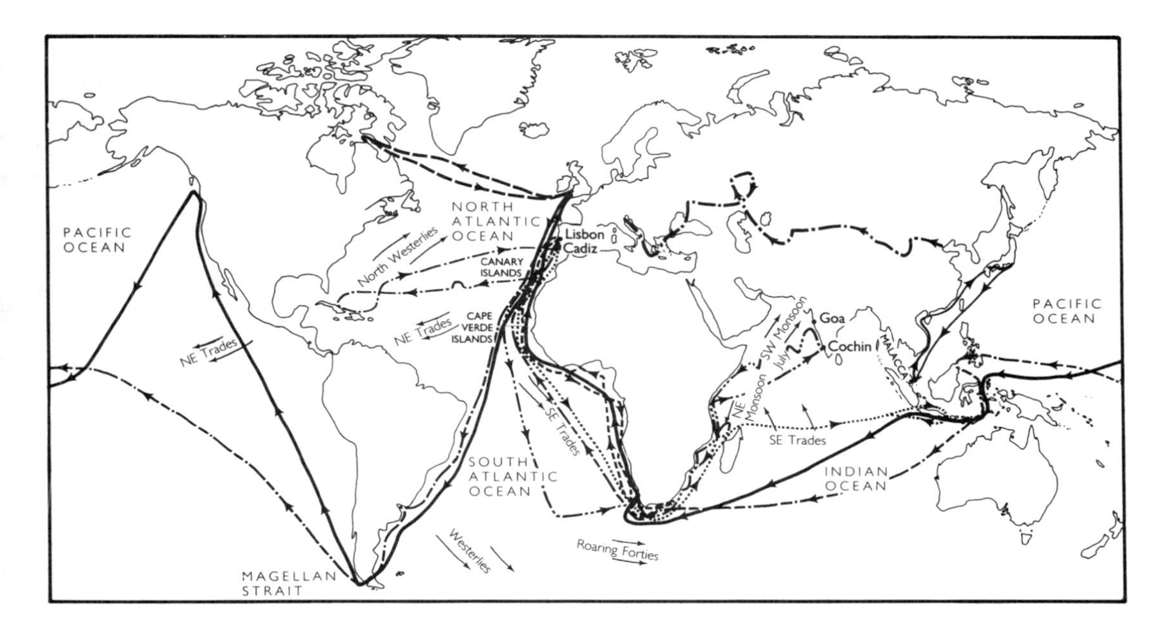

Source 1.1 The Magellan expedition in the Pacific, 1520

Wednesday, the twenty-eighth of November, 1520, we came forth out of the said strait, and entered into the Pacific sea, where we remained three months and twenty days without taking in provisions or other refreshments, and we only ate old biscuit reduced to powder, and full of grubs, and stinking from the dirt which the rats had made on it when eating the good biscuit, and we drank water that was yellow and stinking. We also ate the ox hides which were under the main-yard, so that the yard should not break the rigging: they were very hard on account of the sun, rain, and wind, and we left them for four or five days in the sea, and then we put them a little on the embers, and so ate them; also the saw-dust of wood, and rats which cost half-a-crown each, moreover enough of them were not to be got. Besides the above-named evils, this misfortune which I will mention was the worst, it was that the upper and lower gums of most of our men grew so much that they could not eat, and in this way so many suffered, that nineteen died, and the other giant, and an Indian from the county of Verzin. Besides those who died, twenty-five or thirty fell ill of divers sicknesses, both in the arms and legs, and other places, in such manner that very few remained healthy. However, thanks be to the Lord, I had no sickness. During those three months and twenty days we went in an open sea, while we ran fully four thousand leagues in the Pacific sea.

G Kish (Ed), p. 320.

Question • What concrete results flowed from Magellan's expedition?

Source 1.2 The *Edward Bonaventure* struck by lightning, 1591

Six days after our sending back for England of the *Merchant Royal*, our admiral Captain Raimond in the *Penelope*, and Mr James Lancaster in the *Edward Bonaventure*, set forward to double [round] the Cape of Buona Esperanza, which they did very speedily. The 14 of September we were encountered with a mighty storm and extreme gusts of wind, wherein we lost our general's company, and could never hear of him nor his ship any more, though we did our best endeavour to seek him up and down for a long while. Four days after this, in the morning toward ten of the clock we had a terrible clap of thunder, which slew four of our men outright, their necks being wrung in sunder without speaking any word, and of 94 men there was not one untouched whereof some were stricken blind, others were bruised in their legs and arms, and others in their breasts, so that they voided blood two days after, others were drawn out at length as though they had been racked. But (God be thanked) they all recovered saving only the four which were slain outright. Also with the same thunder our main mast was torn very grievously from the head to the deck, and some of the spikes that were ten inches into the timber, were melted with the extreme heat thereof.

R Hakluyt, p. 362.

Question • Why is Lancaster's voyage considered a landmark in the history of English seafaring?

Source 1.3 A comment by Holden Furber

For an understanding of their view of Asia, it must always be remembered that the Europeans had not experienced the industrial, 'communications,' and medical revolutions of the nineteenth century. We can probably appreciate more easily the great changes in transport and in medicine than those which the industrial revolution brought in Europe's relations with Asia. The monsoon winds determined the timing of the voyages; the time-lag of from five to six months in communication with Asia and of at least a year in receiving a reply to a letter caused difficulties. Contemporaries took these conditions for granted, just as they did the omnipresence and capriciousness of death. They recognized that life expectancy in the tropics was somewhat less than in Europe, but, for many of them, either the possible rewards offset the risks, or the life awaiting them east of the Cape was more appealing than the one facing them at home. Many Europeans of humble social origins preferred the blue skies, warm breezes, and other attractions of the tropics to the squalor to which they were condemned in Europe, and many more, plied with drink and dumped aboard an East Indiaman, had no choice in the matter. Assuredly, among Europeans who developed immunity to tropical diseases in Asia there were many who would have succumbed to pneumonia, tuberculosis, cholera, plague, and other scourges which ravaged Europe before the advent of modern medicine.

H Furber, pp. 6–7.

Question

- Why might this interpretation by Holden Furber be considered '**revisionist**'?

revisionist policy of revision or modification

Vasco da Gama (*c* 1469–1524).

The Might of Islam

Surrounded by the might of Arabic Islam, the Christian States of late medieval Europe, led by Spain and Portugal, were keen to forge alliances with friendly countries in Asia and Africa. In particular, they hoped to make contact with a fabulous ruler called 'Prester John', whose kingdom was thought to lie somewhere in India. In fact there was no such king. But the legend flourished, fuelled by the appearance, in 1165, of a letter supposedly written by Prester John and addressed to Manuel I of Byzantium (source 1.4).

Thus, when Vasco da Gama arrived in 1498 at the Indian town of Calicut, the first European to reach there by sea, he fully expected to meet the 'king of the Christians'. On the first day, his hosts gave him a guided tour of the city, and he mistook the local Kali temple for a church. Our account of this odd encounter comes from the log-book of da Gama's flagship, the *San Rafael*, which is reputed to have been kept by one of his officers, Alvaro Velho (source 1.5).

Source 1.4 The legend of 'Prester John', *c.* 1165

John, Priest by the Almighty power of God and the strength of Our Lord Jesus Christ, King of Kings and Lord of Lords, to his friend Emmanuel, Prince of Constantinople, greeting, wishing him health and the continued enjoyment of the Divine Favour.

It hath been reported to our Majesty that thou holdest our Excellency in esteem, and that the knowledge of our highness has reached thee . . .

If thou shouldst have any desire to come into the kingdom of our majesty, we will place thee in the highest and most dignified position in our household, and thou mayest abundantly partake of all that pertains to us. Shouldst thou desire to return, thou shalt go laden with treasures. If indeed thou desirest to know wherein consists our great power, then believe without hesitation, that I, Prester John, who reign supreme, surpass in virtue, riches and power all creatures under heaven. Seventy kings are our tributaries. I am a zealous Christian and universally protect the Christians of our empire, supporting them by our alms. We have determined to visit the sepulchre [that is, Jerusalem] of our Lord with a very large army, in accordance with the glory of our majesty to humble and chastise the enemies of the cross of Christ and to exalt his blessed name.

Honey flows in our land, and milk everywhere abounds. In one region there no poison exists and no noisy frog croaks, no scorpions are there, and no serpents creeping in the grass.

No venomous reptiles can exist there or use there their deadly power. In one of the heathen provinces flows a river called the Indus, which, issuing from Paradise, extends its windings by various channels through all the province; and in it are found emeralds, sapphires, carbuncles, topazes, chrysolites, onyxes, beryls, sardonyxes, and many other precious stones.

A P Newton, pp. 174–7.

Questions

- What is the meaning of the 'c.' before the date?
- Given that this letter was a forgery, is it a valid historical source?

Source 1.5 Vasco da Gama at the Court of the Zamorin of Calicut, 1498

On the following morning, which was Monday, 28th May, the captain-major set out to speak to the king, and took with him thirteen men, of whom I was one. We put on our best attire, placed bombards in our boats, and took with us trumpets and many flags . . .

When we arrived they took us to a large church, and this is what we saw:

The body of the church is as large as a monastery, all built of hewn stone and covered with tiles. At the main entrance rises a pillar of bronze as high as a mast, on the top of which was perched a bird, apparently a cock. In addition to this, there was another pillar as high as a man, and very stout. In the centre of the body of the church rose a chapel, all built of hewn stone, with a bronze door sufficiently wide for a man to pass, and stone steps leading up to it. Within this sanctuary stood a small image which they said represented Our Lady. Along the walls, by the main entrance, hung seven small bells. In this church the captain-major said his prayers, and we with him.

We did not go within the chapel, for it is the custom that only certain servants of the church, called *quafees*, should enter. These *quafees* wore some threads passing over the left shoulder and under the right arm, in the same manner as our deacons wear the stole. They threw holy water over us, and gave us some white earth, which the Christians of this country are in the habit of putting on their foreheads, breasts, around the neck, and on the forearms. They threw holy water upon the captain-major and gave him some of the earth, which he gave in charge of someone giving them to understand that he would put it on later.

Many other saints were painted on the walls of the church, wearing crowns. They were painted variously, with teeth protruding an inch from the mouth, and four or five arms.

After we had left that place, [we were taken to the palace and ushered into the presence of the king]. . .

And the captain told him he was the ambassador of a king of Portugal, who was lord of many countries and the possessor of great wealth of every description, exceeding that of any king of these parts; that for a period of sixty years his ancestors had annually sent out vessels to make discoveries in the direction of India, as they knew that they were Christian kings there like themselves. This, he said, was the reason which induced them to order this country to be discovered, not because they sought for gold or silver, for of this they had such abundance that they needed not what was to be found in this country. He further stated that the captains sent out travelled for a year or two, until their provisions were exhausted, and then returned to Portugal, without having succeeded in making the desired discovery. There reigned a king now whose name was Dom Manuel, who had ordered him to build three vessels, of which he had been appointed captain-major, and who had ordered him not to return to Portugal until he should have discovered the king of the Christians, on pain of having his head cut off. That two letters had been entrusted to him to be presented in case he succeeded in discovering him, and that he would do so on the ensuing day; and, finally, he had been instructed to say by word of mouth that he [the king of Portugal] desired to be his friend and brother.

C D Ley (Ed), pp. 29–33.

Question • Did da Gama's party actually visit a Christian church? If not, what manner of building was it?

The Mysterious East

Despite the hoaxes, like the Prester John letter and the book of *Travels* made up by Sir John de Mandeville (d. 1372), Europeans received some genuine information about Asia, the result of overland journeys via the central Asian 'silk road' to China, which was reopened to traffic by the Mongols during the 13th century. Perhaps the most famous of these accounts was that by Marco Polo (1254–1324?), a Venetian merchant. With his father, Nicolo, he spent 17 years at the court of Kublai Khan. After his return to Europe in 1295, Marco enlisted as a soldier in the Venetian army fighting Genoa and was taken captive. While imprisoned, he dictated his life story to a fellow prisoner, and it subsequently became very popular. Our excerpt (source 1.6) shows why.

However, it was not only Asia's enormous wealth and power that appealed to the European mind. Then, as now, Europeans were attracted by Asia's exotic cults and mysterious customs, by its teeming cities and wild jungles. Source 1.7, taken from a short story by the 20th century English writer Somerset Maugham, who was a frequent visitor to South-East Asia, describes vividly the lure of the 'mystic East'.

Source 1.6 Marco Polo on the city of 'Cambaluc', China, 1298

You must know that the city of Cambaluc hath such a multitude of houses, and such a vast population inside the walls and outside, that it seems quite past all possibility. There is a suburb outside each of the gates, which are twelve in number; and these suburbs are so great that they contain more people than the city itself [for the suburb of one gate spreads in width till it meets the suburb of the next, whilst they extend in length some three or four miles]. In those suburbs lodge the foreign merchants and travellers, of whom there are always great numbers who have come to bring presents to the Emperor, or to sell articles at Court, or because the city affords so good a mart to attract traders. [There are in each of the suburbs, to a distance of a mile from the city, numerous fine hostelries for the lodgment of merchants from different parts of the world, and a special hostelry is assigned to each description of people, as if we should say there is one for the Lombards, another for the Germans, and a third for the Frenchmen.] And thus there are as many good houses outside of the city as inside, without counting those that belong to the great lords and barons, which are very numerous . . .

Moreover, no public woman resides inside the city, but all such abide outside in the suburbs. And 'tis wonderful what a vast number of these there are for the foreigners; it is a certain fact that there are more than 20,000 of them living by prostitution. And that so many can live in this way will show you how vast is the population . . .

To this city also are brought articles of greater cost and rarity, and in greater abundance of all kinds, than to any other city in the world. For people of every description, and from every region, bring things (including all the costly wares of India, as well as the fine and precious goods of Cathay itself with its provinces), some for the sovereign, some for the court, some for the city which is so great, some for the crowds of Barons and Knights, some for the great hosts of the Emperor which are quartered round about; and thus between court and city the quantity brought in is endless.

G Kish (Ed), pp. 251–3.

Question • Why do you think 'Polo's Book' immediately became widely read throughout Europe?

Source 1.7 'Neil MacAdam' arrives in Malaya

Now they were steaming up the river. At the mouth was a straggling fishermen's village standing on piles in the water; on the bank grew thickly nipah palm and the tortured mangrove; beyond stretched the dense green of the virgin forest. In the distance, darkly silhouetted against the blue sky, was the rugged outline of a mountain. Neil, his heart beating with the excitement that possessed him, devoured the scene with eager eyes. He was surprised. He knew his Conrad almost by heart and was expecting a land of brooding mystery. He was not prepared for the blue milky sky. Little white clouds on the horizon, like sailing boats becalmed, shone in the sun. Here and there, on the banks, were Malay houses with thatched roofs, and they nestled cosily among fruit trees. Natives in dug-outs rowed, standing, up the river. Neil had no feeling of being shut in, nor, in that radiant morning, of gloom, but of space and freedom. The country offered him a gracious welcome. He knew he was going to be happy in it.

W S Maugham, p. 439.

Question • What other attractions, besides those described, might a visitor find in South-East Asia?

The Colonial Lifestyle

Life in Asia was dangerous and often uncomfortable, but there were compensations. As source 1.8 reveals, Europeans living in the East in the early period drank heavily, partly due to

Master and servants in Bangalore, India, late 19th century.

discontent and partly because hard liquor was regarded as a good preventive measure against disease. Later, as the colonial settlements expanded, social and sporting clubs were formed and, with domestic servants plentiful and cheap, 'opulent dinners and suppers'—such as those observed in the Shanghai International Settlement by the German traveller A H Exner, quoted by Franke (source 1.9)—were common events.

However, it was in the temperate 'hill stations', such as Simla, Ootacamund and Darjeeling in India and Nuwara Elliya in Ceylon, that the colonial lifestyle reached its leisurely peak. In these high-altitude outposts, a combination of cool climate and open spaces allowed the Europeans to create environments remarkably like those they had left behind at home, and in the process, to forget, at least for a time, the heavy burdens that awaited them in the dusty, crowded plains.

Source 1.8 Possessions found in the house of Major James Kilpatrick, Bengal Army, at the time of his decease, 8 November 1757

4 pairs of scarlet breeches (trousers)
4 pairs of black breeches
47 pairs of gingham breeches
4 hats (with gold and silver lace trimmings)
97 pairs of stockings
58 old shirts
161 new shirts
37 neckcloths
3 pairs of long drawers
79 waistcoats
3 quilted Banyan coats
42 handkerchiefs
A set of musketo curtains
12 pairs of sneakers
15 curry plates
Many gold and silver ornaments
97 bottles of Madeira wine
116 bottles of Orange Shrub [a kind of liqueur]
48 bottles of Mango Shrub
13 dozen bottles of brown Arrack
8 dozen bottles of white Arrack
28 bottles of Goa Arrack
A barrel containing 115 gallons of Batavia Arrack
A hogshead of English beer
925 empty bottles
A dried salmon

P Spear, pp. 181–2.

Question • What can one deduce from this list about the lifestyle of the British in 18th century India?

Source 1.9 Life in the international settlement, Shanghai, *c.* 1889

. . . Apart from this business activity, which only requires our attention now and again, we spend our time in the way of life which is usual here, which must be described as on the whole exceedingly monotonous. . . . The comparatively few Europeans who live here have no one else to turn to but themselves in their social intercourse . . . and thus as a rule each day goes by like all the others. In the early morning an hour or two is spent at work, and in the afternoon, when the heat permits, one can play lawn tennis with the few ladies in the colony, and it is possible to take a ride later on into the monotonous area of the town nearby or perhaps beyond the mud wall of the city to the settlement racecourse, which lies in the open country and is completely without shade. Toward mid-day one meets the other men in the club for whisky and soda, and to hear the gossip of the town, and in the evening there will be a game of billiards or skittles in the same place. . . . Since the 'real' work is usually restricted for Europeans, as in most places overseas, to a few days in the week, that is, to the days round about the arrival and departure of the mail steamer, they have an extraordinary amount of free time left at their disposal, which in the absence of any opportunity for intellectual amusement, tends largely to be spent in entertaining and being entertained at extraordinarily opulent dinners and suppers.

W Franke, pp. 87–8.

Question • On the basis of this description, would you have wanted to be a European resident of Shanghai in the later 19th century?

Fortune Hunters

indigo plant from which special blue dye (also called indigo) is obtained

Asia was once the global leader in science and technology. It was also once the producer of the world's finest manufactures, whose equal would not be seen in the West until the Industrial Revolution. Persia, for example, was even then world-famous for its carpets, as the English lawyer and scholar Richard Hakluyt reminded his friend, the dyer Morgan Hubblethorn, when the latter sought his advice in 1579 (source 1.10). Asia also supplied Europe with much-needed (and hence expensive) tropical and subtropical agricultural products, such as spices, **indigo**, coffee and tea; and in source 1.12, taken from one of his best-known novels, *Lord Jim*, Joseph Conrad (1857–1924) offers us a glimpse into the profitable pepper trade of the Indies. Greed, he demonstrates, lay at the heart of this commerce.

English merchants being tortured by the Dutch on Amboyna, 1623. The prisoner, strapped to a rack, is being simultaneously drenched in water and burned under the armpits.

But trade was not the only avenue to wealth that Asia offered. By the 18th century, the European East India Companies operating in Java and India were substantial military powers, and local rulers were prepared to pay large sums for their support. For example, in 1757 the English East India Company's Robert Clive (1725–74) engineered a palace coup in Bengal and became, virtually overnight, the richest man in England (source 1.11).

Later still, Asian colonies generated jobs for the boys, prompting John Stuart Mill to remark that the British Empire was a 'vast system of outdoor relief for the English aristocracy'.[6] Although the salaries attached to them seem modest (source 1.13), the cost of living was also low. In 1850, very comfortable lodgings could be obtained in London for 48 shillings a weeks, while in the 1830s, 400 pounds per annum could buy a house, two maids, a horse and a groom.[7]

Source 1.10 Richard Hakluyt commends Persian carpets, 1579

In Persia you shall find carpets the best of the world, and excellently coloured: you must use means to learn all the order of the dyeing which are so dyed as neither rain, wine, nor yet vinegar can stain.

For that in Persia they have great colouring of silks, it behoves you to learn that also, for that cloth dyeing and silk dyeing have a certain affinity, and your merchants mind to bring much raw silk into the realm, and therefore it is more requisite you learn the same.

In Persia there are that stain linen cloth: it is not amiss you learn it if you can: it hath been an old trade in England, whereof some excellent cloths yet remain: but the art is now lost, and not to be found in the realm.

R Hakluyt, p. 207.

Question • What other articles manufactured in Asia were in high demand in pre-modern Europe?

Source 1.11 Sums distributed to senior officials of the English East India Company by Nawab Mir Jafar on the occasion of his elevation to the throne of Bengal, 1757

	Rs.	Rs.	£
Mr. Drake (Governor)		280,000	31,500
Colonel Clive as Second in the Select Committee	2,80,000		
Ditto as Commander-in-Chief	2,00,000		
Ditto as a private donation	16,00,000		
		20,80,000	234,000
Mr. Watts as a member of the Committee	240,000		
Ditto as a private donation	800,000		
		1040,000	117,000
Major Kilpatrick		240,000	27,000
Ditto as a private donation		300,000	33,750
Mr. Maningham		240,000	27,000
Mr. Becher		240,000	27,000
Six members of Council one *lack* each		600,000	68,200
Mr. Walsh		500,000	56,250
Mr. Scrafton		200,000	22,500
Mr. Lushington		50,000	5,625
Captain Grant		100,000	11,250
Stipulation to the navy and army			600,000
			1,261,075

S C Hill, p. 395.

Question • Why did Robert Clive get so much more than anyone else?

Source 1.12 The pepper traders of 'Patusan'

The seventeenth-century traders went there for pepper, because the passion for pepper seemed to burn like a flame of love in the breast of Dutch and English adventures about the time of James the First. Where wouldn't they go for pepper! For a bag of pepper they would cut each other's throats without hesitation, and would forswear their souls, of which they were so careful otherwise: the bizarre obstinacy of that desire made them defy death in a thousand shapes; the unknown seas, the loathsome and strange diseases; wounds, captivity, hunger, pestilence, and despair. It made them great! By heavens! it made them heroic; and it made them pathetic, too, in their craving for trade with the inflexible death levying its toll on young and old. It seems impossible to believe that mere greed could hold men to such a steadfastness of purpose, to such a blind persistence in endeavour and sacrifice. And indeed those who adventured their persons and lives risked all they had for a slender reward. They left their bones to lie bleaching on distant shores, so that wealth might flow to the living at home. To us, their less tried successors, they appear magnified, not as agents of trade but as instruments of a recorded destiny, pushing out into the unknown in obedience to an inward voice, to an impulse beating in the blood, to a dream of the future. They were wonderful; and it must be owned they were ready for the wonderful. They recorded it complacently in their sufferings, in the aspect of the seas, in the customs of strange nations, in the glory of splendid rulers.

J Conrad, pp. 226–7.

Question • Why did pepper fetch such high prices?

Source 1.13 Establishment of the Ceylon Civil Service, 1845

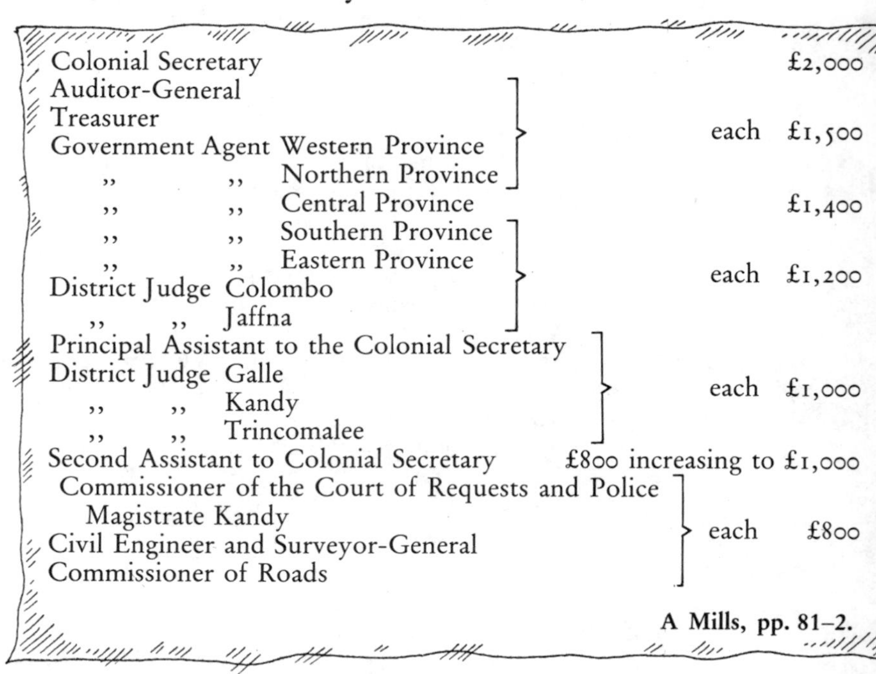

Colonial Secretary	£2,000
Auditor-General	each £1,500
Treasurer	each £1,500
Government Agent Western Province	each £1,500
,, ,, Northern Province	each £1,500
,, ,, Central Province	£1,400
,, ,, Southern Province	each £1,200
,, ,, Eastern Province	each £1,200
District Judge Colombo	each £1,200
,, ,, Jaffna	each £1,200
Principal Assistant to the Colonial Secretary	each £1,000
District Judge Galle	each £1,000
,, ,, Kandy	each £1,000
,, ,, Trincomalee	each £1,000
Second Assistant to Colonial Secretary	£800 increasing to £1,000
Commissioner of the Court of Requests and Police Magistrate Kandy	each £800
Civil Engineer and Surveyor-General	each £800
Commissioner of Roads	each £800

A Mills, pp. 81–2.

Question • How long might it take a civil servant to rise to the rank of judge or agent?

The Christian Mission

The current trend among historians is to reduce the importance of religion as a causative factor. In this respect, Carlo Cipolla's comment quoted earlier—that religion was the pretext for European expansion and 'gold' the motive—may be taken as typical. However, it would be a mistake to dismiss religious enthusiasm as a mere cloak for other concerns.

The early modern period in Europe was a devout, even fanatical, age that was dominated by the religious struggles of the Reformation and Counter-Reformation. Even in the 19th century, Europeans were typically avid churchgoers and Bible-readers. Moreover, the men (and later women) who went to Asia as missionaries were, by definition, people of intense religious conviction. When Alexander Duff was chosen in 1828 by the St Andrew's Missionary Society of Edinburgh to serve in India, he confided to his friend and teacher, Dr Chalmers, that he was 'almost transported with joy' at the prospect 'of winning souls for Christ'—even though he fully expected the work to involve personal sacrifice and hardship (source 1.15).

Jesuit member of the Society of Jesus, a Catholic religious order founded by Ignatius Loyola

martyrdom sufferings and death for a great cause, usually religious

Likewise, the Basque **Jesuit** missionary, Francis Xavier, knew that he was risking his life when he landed at the port of Kagoshima bent on converting Buddhist Japan. Yet, as he observes in a letter of 1549 to his fellow Jesuits at Goa, written shortly after his arrival (source 1.14), death held no fears for him, since **martyrdom** at the hands of the Buddhist priests (*bonzes*) would merely bring closer the moment of his own salvation. Can we doubt their sincerity?

Source 1.14 Francis Xavier: What we do in Japan, 1549

What we in these parts endeavor to do, is to bring people to the knowledge of their Creator and Saviour, Jesus Christ Our Lord. We live with great hope and trust in Him to give us strength, grace, help, and favor to prosecute this work. It does not seem to me th the laity will oppose or prosecute us of their own volition, but only if they are importuned [forced] by the Bonzes. We do not seek to quarrel with them, neither for fear of them will we cease to speak of the glory of God, and of the salvation of souls. They cannot do us more harm than God permits, and what harm comes to us through them is a mercy of God; if through them for His love, service, and zeal for souls, He sees good to shorten the days of our life, in order to end this continual death in which we live, our desires will be speedily fulfilled and we will go to reign forever with Christ. Our intentions are to proclaim and maintain the truth, however much they may contradict us, since God compels us to seek rather the salvation of our future than the safety of our present lives; we endeavoring with the grace and favor of Our Lord to fulfil this precept, He giving us internal strength to manifest the same before the innumerable idolatries which there are in Japan.

C R Boxer, *Christian Century*, p. 404.

Question • What gave missionaries like Xavier the audacity to preach the Gospel in remote and hostile places?

Source 1.15 Alexander Duff decides to become a missionary in India, 1828

I am now prepared to reply to the Committee in the words of the Prophet, 'Here am I, send me.' The work is most arduous, but is of God, and must prosper; many sacrifices painful to 'flesh and blood' must be made, but not any correspondent to the glory of winning souls for Christ. With the thought of this glory I feel myself almost transported with joy, everything else appears to fall out of view as vain and insignificant. The kings and great men of the earth have reared the sculptured monument and the lofty pyramid with the vain hope of transmitting their names with reverence to succeeding generations; and yet the sculptured monument and the lofty pyramid do crumble unto decay, and must finally be burnt up in the general wreck of dissolving nature; but he who has been the means of subduing one soul to the Cross of Christ, hath reared a far more enduring monument—a monument that will outlast all time, and survive the widespread ruins of ten thousand worlds; a trophy which is destined to bloom and flourish in immortal youth in the land of immortality.

W Paton, pp. 50–1.

Question • Why were missionaries not discouraged by their failure to achieve masses of converts?

'Our German Missionary'; watercolour by Captain G F Atkinson. The tactic of street-preaching, which Protestant missionaries employed, was quite foreign to Asia and offended pious natives (see source 6.14).

The Call of Duty

Many colonial servants were attracted by the good salaries which, as we have seen, were attached to posts in the upper levels of the administration. However, others were motivated by the desire to improve the condition of those whom Rudyard Kipling called 'the lesser breeds without the law'. More simply, they were motivated

noblesse oblige privilege entails responsibility

by the desire to serve their country overseas. Indeed, many families had a tradition of such service based on the old upper class ethic of ***noblesse oblige***.

Lord Hardinge, who served as Viceroy of India during the period 1910–16, recalls in his memoirs that he was encouraged, from an early age, to 'follow in the footsteps of my grandfather', who had been Governor-General in the 1840s.[8] And in source 1.16 a similar Indian connection in the geneological history of the literary Thackeray family can be traced.

The public schools also played an important role, particularly in Britain, as nurseries of imperialism. George Nathaniel (later Lord) Curzon (1859–1925), first heard about the grandeur of the British Raj while a schoolboy at Eton, listening to a talk by a former Law Member of the Government of India, Sir James Fitzjames Stephen (source 1.17). The fascination for the subcontinent he acquired then and there never left him, and he went on to become one of the greatest of India's proconsuls (governors).

Source 1.16 The Thackeray family tree, 1693–1863

Rev. Dr Thomas Thackeray 1693–1797 m. in 1729 Anne, a daughter of John Woodward

W M Thackeray (1749–1813), B.C.S. m. in 1776 Amelia, daughter of Lt. Col. Richmond Webb of the military service

Henrietta m. in 1771 James Harris (b?–1790), B.C.S.

Jane m. in 1772 James Rennell, (1742–1830), B.M.S.

William (1778–1823), Madras C.S.

Webb (1790–1807), Madras C.S.

St. John (1791–1824), Madras C.S.

Charles (1794–1846), Bar-at-law at Calcutta.

Thomas (1789–1814), Madras M.S.

Emily m. in 1803 J T Shakespeare (1783–1825), B.C.S.

Augusta m. in 1816 John Elliot (1765–1818), B.C.S.

Rev. Francis (1793–1842) m. in 1829 Mary Anne, daughter of John Shakespeare, B.C.S.

Richmond (1781–1815), B.C.S. m. in 1810 Anne, daughter of J H Becher, B.C.S.

Sir E T Thackeray 1836–d?) of the military service m. in 1862 Amy Mary Anne, second daughter of Eyre Evans Crome.

W M Thackeray, the novelist, (1811–1863) m. in 1837 Harriet, sister of Capt. Shawe of the military service.

S C Ghosh, p. 177.

Question • How many generations of Thackerays had served in India by the mid-19th century?

Source 1.17 George Curzon's introduction to India, *c.* 1872

Sir James Stephen came down to Eton and told the boys that listened to him, of whom I was one, that there was in the Asian continent an empire more populous, more amazing, and more beneficent than that of Rome; that the rulers of that great dominion were drawn from the men of our own people; that some of them might perhaps in the future be taken from the ranks of the boys who were listening to his words. Ever since that day, and still more since my first visit to India in 1887, the fascination and, if I may say so, the sacredness of India have grown upon me, until I have come to think that it is the highest honour that can be placed upon any subject of the Queen that in any capacity, high or low, he should devote such energies as he may possess to its service.

K Rose, pp. 199–200.

Question • Why would a boy at Eton be impressed by the comparison of the British Indian Empire with that of Rome?

Mounted on an elephant, Lord Curzon (1859–1925) processes through Delhi during the Darbar of 1903.

2 From Commerce to Conquest

We seem, as it were, to have conquered and peopled half the world in a fit of absence of mind!

Sir John Seeley, 1883

THIS CHAPTER FOCUSES ON the second major phase of European expansion which peaked in the so-called 'New Imperialism' of the later 19th century. In particular, it seeks to explain why the European powers generally abandoned their previous commitment to 'peaceful' trade and proselytisation (religious conversion) in favour of a program of conquest and annexation. What political, economic or personal ends were serve by the imposition of direct colonial rule on non-European people

KEY DATES

1740 Beginning of the Carnatic Wars (1740–56), which saw a struggle between Britain and France for supremacy in Sout India.

1784 British Parliament enacts Pitt's India Act, in an attempt to restrain the independence of the English East India Company.

1793 Earl Macartney (1737–1806) leads a British embassy to the court of the Ch'ing emperor Ch'ien-lung (*r.* 1736–95), in an unsuccessful bid to obtain extended trading rights in China.

1819 (Sir) Stamford Raffles purchases the island of Singapore from the Sultan of Johore on behalf of the English East India Company.

1841 (Sir) James Brooke (1803–68) installed by the Sultan of Brunei as Governor of Sarawak (Borneo).

1842 China, humiliated by Britain in the First 'Opium War' (1839–42), forced to surrender Hong Kong and open five ports to trade under the Treaty of Nanking (Nanjing).

1859 Publication of Charles Darwin's *Origin of Species* leads to speculation about racial 'evolution'.

1860 French expedition captures Saigon (Ho Chi Minh City), annexes four provinces of Cochin-China (Vietnam).

1869 Suez Canal opened, dramatically shortening the length of the sea-voyage from Europe to Asia.

1885 British capture Mandalay, and depose King Theebaw; uppe Burma is absorbed into the Indian Empire.

1889 First *Exposition Coloniale* (Colonial Exhibition) held in Paris.

1899 United States President William McKinley (1843–1901) annexes the Philippines.

ISSUES AND PROBLEMS

European imperialism in Asia underwent substantial changes during the 17th and 18th centuries. First, the 16th century saw th advent of several new players in the imperial game: notably the

English and the Dutch. In 1591, Captain James Lancaster, sailing via the Cape route (source 1.2), penetrated as far as Malaya, and four years later Cornelis Houtman took his rebellious crew all the way to Bantam on the north-west coast of Java. By 1598, 22 ships had left Dutch ports for the East and a further 43 followed between 1598 and 1601. The Portuguese did not take kindly to this Protestant intrusion, which threatened their lucrative monopoly of the spice trade. They resisted the intruders fiercely, the garrison at Colombo only surrendering to the Netherlanders after an eight month siege in which all but 73 defenders died from starvation and disease. However, the power of the newcomers proved irresistible, and by 1660 virtually all the Portuguese bases in South-east Asia were in Dutch hands.

Second, this new wave of expansion was very largely sponsored by private enterprise. Whereas the Spanish and Portuguese expeditions to the East had been directed and financed by the State, those of the northern powers were carried out by licensed or 'chartered' companies—the English East India Company, the Ostende Company, the Amsterdam-based Vereenigde Oost-Indische Compagnie (VOC), the French Compagnie des Indes, and the East India Company of Copenhagen. Using the new **joint stock** mechanism, these companies were able to amass very large amounts of working capital, which enabled them to compete effectively in Asia, despite heavy losses from shipwreck. The English Company's 1617–32 share float, for example, raised £1.5 million. Yet the spice, coffee, textile and indigo trades in this period were so profitable that the VOC was able to pay its shareholders dividends of 18 per cent and more throughout the century.

joint stock commercial enterprise where 'stock' is held by proprietors or 'stockholders'

Most importantly, the nature of imperialism changed as the Europeans in the East became steadily more powerful and more ambitious. The Portuguese, while supreme at sea, never managed to carve out a substantial land empire in Asia. They did not see the need for one, since neither trade nor proselytisation required the ownership or control of large amounts of territory. At first, the English and Dutch followed a similar non-interventionist policy. As England's first ambassador to the Mughal Empire, Sir Thomas Roe, told the English Company's employees at Surat in 1612: 'if you will profit, seek it at sea and in quiet trade. For it is an error to affect [wage] land wars in India'.[1] But this cautious approach was slowly abandoned. In 1619, the Dutch fortified Batavia, their Indonesian headquarters, and in 1641 the English did the same in Madras. In 1651, Dutch Governor-General Cornelis Speelman attacked and defeated the Sultan of Makassar, and in 1678 sent troops to the aid of the Javanese dynasty of Mataram, which was threatened by a rebellion. In 1686, the English declared war on the mighty Mughal Empire. In the 1740s the English and French Companies became militarily involved (on opposite sides) in succession struggles for the south Indian thrones of Hyderabad and the Carnatic. By the end of the 18th century, most of eastern India and Java were effectively under European control.

What brought about these changes? Portugal was only a small, agricultural country. Lacking manpower, sophisticated financial institutions, and a metallurgy industry (essential for the casting of cannon), it simply did not have the resources to compete effectively, on a global scale, with England and The Netherlands. Although equally small, these countries were more populous and were thriving commercially. However, the transition from commerce and preaching to conquest and territorial annexation seems, at first sight, paradoxical. Warfare was expensive, disruptive and reduced profits (see chapter 9). Furthermore, as in the case of Japan, political meddling tended to harm the cause of Christian conversion. In addition, conquest was undertaken, in most cases, against the explicit wishes of the metropolitan authorities (source 2.18), who, as the British Prime Minister, Lord Shelburne, made clear in a speech of 1783, continued to 'prefer commerce to dominion'.[2] Why did the local officials of the East India Companies embrace tactics that were likely to be counter-productive? And why were they not pulled into line by their governments at home?

THE DEBATE

Wrestling with these problems in the late 19th century, the eminent historian, Sir John Seeley, concluded that expansion must have happened 'blindly', by 'accident'.[3] However, even if the conquest of Asia was unplanned—and here, Seeley probably has a valid point—it was definitely not accidental. Annexations do not happen by themselves. In every case someone, somewhere, was responsible. But who? Why?

There are currently four main theories. The first emphasises the factor of *prestige*. According to this theory, the emergence during the 1870s of Italy and Germany—two powerful and aggressive **nation-states**—upset the European balance of power. This forced Britain and France to seek to rebuild their declining fortunes by adding to their overseas empires. Professor John F Cady has claimed, 'The taproot of French imperialism . . . from first to last was national pride—pride of culture, reputation, prestige and influence'.[4]

The second main line of explanation, pioneered by J A Hobson, stresses *economic imperatives*—the need for the industrialised countries to find outlets for the products of their **factories** and the desire of investors to reap higher returns for their 'surplus' capital.

The third theory, however, maintains that the Europeans were

nation-states states with populations drawn mainly from a single ethnic group

factories specifically warehouses; generally European settlements in Asian ports during 16th–17th centuries

vilising mission 9th century doctrine hat Europeans should pread their 'superior' ulture among backward' peoples

urbulent frontier oncept that urbulent' onditions in Asia romoted expansion s colonies sought ecure frontiers

ecentralised dministrative affairs istributed among ocal centres instead f all power being entred in only one lace

ingoism late 19th entury term for mperialistic war-mongering

not so much out to take as to give; and that Asia was colonised with the intention of implanting Western technology and values. Wherever a wandering Westerner happened across a 'backward' people, writes Pakistani historian K K Aziz, 'it became his duty to conquer them, absorb them and make them a part of the civilised and civilising machinery of which he was in charge'.[5] This argument could be described as the **civilising mission**.

Finally, there is the **turbulent frontier** explanation. The essence of this theory, developed by F G Hutchins, is that, in a period when communications were slow and unreliable, imperial decision-making had to be **decentralised**, which left key decisions about offence and defence in the hands of ambitious officials on the spot. Such men, it is argued, were predisposed to expansion by motives of self-interest (fighting a successful frontier campaign being a guarantee of promotion) and by the logic that, in an ever-changing world, the only secure frontiers were natural ones, such as oceans and mountains. This is central to the paradox which puzzled Professor Seeley.

Were the Europeans Reluctant Imperialists?

Did the Europeans actively seek empire in Asia, or was the responsibility of administration thrust upon them, an unwelcome but irresistible burden? To judge from the speeches and actions of the home authorities, conquest and annexation of territory was not, at least at the beginning, a fundamental part of their plans for Asia. Indeed, there is considerable evidence that the European powers in the late 18th and early 19th centuries were actually against overseas entanglements. For example, the India Act was pushed through the British Parliament in 1784 by the 25 year old leader of the Tory Party William Pitt (first Earl of Chatham, 1759–1806). This Act appeared to make such activities, if pursued with a deliberately aggressive purpose, illegal (source 2.1).

Similarly, the mid-19th century has often been portrayed by historians as an era of 'anti-imperialism' in Britain. During this period politicians as diverse as Richard Cobden (source 2.17) and Benjamin Disraeli characterised the colonies as expensive 'millstones' around the neck of the English tax-payer.[6] Later, the Home governments started to change their tune. Yet even during the peak of imperialism after 1870, when national rivalries unleashed a mad 'scramble' for new colonies (source 2.3), and a spirit of militaristic **jingoism** was expressed throughout the land, there were a few radical and left-wing figures who continued to protest against what was happening. One was the English poet and one-time diplomat Wilfred Scawen Blunt (1840–1922), who spent nearly fifty years exposing colonial oppression and publicising the cause of Irish, Egyptian and Indian independence. Source 2.2 comes from a diary entry of December 1900.

Source 2.1 Pitt's India Act, 1784

XXXIV And whereas, to pursue schemes of conquest and extension of dominion in India, are measures repugnant to the wish, the honour, and policy of this nation; be it therefore further enacted by the authority aforesaid, that it shall not be lawful for the Governor General and Council of Fort William aforesaid, without the express command and authority of the said Court of Directors, or of the Secret Committee of the said Court of Directors, in any case, except when hostilities have actually been commenced, or preparations actually made for the commencement of hostilities, against the British nation in India, or against some of the princes or states dependant thereon, or whose territories the said United Company shall be at such time engaged by any subsisting treaty to defend or guaranty either to declare war or commence hostilities, or enter into any treaty for making war, against any of the country princes or states in India, or any treaty for guarantying the possessions of any country princes or states; and that in such case it shall not be lawful for the said Governor General and Council to declare war or commence hostilities, or enter into treaty for making war, against any other prince or state than such as shall be actually committing hostilities, or making preparations as aforesaid, or to make such treaty for guarantying the possessions of any prince or state, but upon the consideration of such prince or state actually engaging to assist the Company against such hostilities commenced, or preparations made as aforesaid; and in all cases where hostilities shall be commenced, or treaty made, the said Governor General and Council shall, by the most expeditious means they can devise, communicate the same unto the said Court of Directors, together with a full state[ment] of the information and intelligence upon which they shall have commenced such hostilities, or made such treaties, and their motives and reasons for the same at large.

P J Marshall, *Problems of Empire*, pp. 167–8.

Question • Did this Act expressly prohibit expansion?

Source 2.2 Wilfred Scawen Blunt, a critic of imperialism, 1900

The old century is very nearly out, and leaves the world in a pretty pass, and the British Empire is playing the devil in it as never an empire before on so large a scale. We may live to see its fall. All the nations of Europe are making the same hell upon earth in China, massacring and pillaging and raping in the captured cities as outrageously as in the Middle Ages. The Emperor of Germany gives the word for slaughter and the Pope looks on and approves. In South Africa our troops are burning farms under Kitchener's command, and the Queen and the two Houses of Parliament and the bench of bishops thank God publicly and vote money for the work. The Americans are spending fifty millions a year on slaughtering the Filipinos; the King of the Belgians has invested his whole fortune on the Congo, where he is brutalizing the negroes to fill his pockets. The French and Italians for the moment are playing a less prominent part in the slaughter, but their inactivity grieves them. The whole white race is revelling openly in violence, as though it had never pretended to be Christian. God's equal curse be on them all! So ends the famous nineteenth century into which we were so proud to have been born.

W S Blunt, p. 464.

Question • Would Blunt's views have appealed to many of his countrymen?

Source 2.3 The growth of European colonial empires 1830–1919

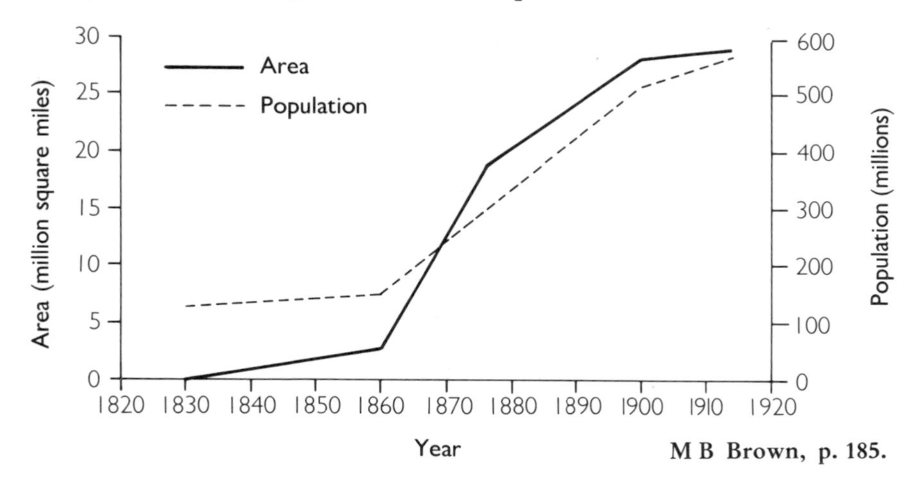

M B Brown, p. 185.

Question

- According to this graph, the pace of European expansion began to accelerate spectacularly during the 1860s. What colonial acquisitions took place in Asia between 1860 and 1875?

Rivalry between European Nations

Nationalism and imperialism are closely related. In the Asian context, as we shall see, nationalism arose partly as a reaction to the oppressive side of European imperialism. In Europe, however, rivalries between nation-states, which increased sharply after the unification of Germany and Italy in 1871, caused otherwise sensible politicians to indulge in a frantic grab for territory simply to beat other nations. Ironically, one of the principal offenders in this regard was the British Liberal Party, which, under the strong leadership of William Gladstone (1809–98), was returned to office in 1868 on a program of 'peace, retrenchment and reform'. In source 2.4, the Liberals' Colonial Secretary, Lord Kimberley, explains to Gladstone why, despite their Party's platform, a forward policy in Malaya must be pursued.

New Imperialism accelerated European expansion at the end of the 19th century

However, while the **New Imperialism** of the late 19th century throve on suspicion and fear, it was not entirely irrational. For France, colonial success was vital for the nation to recover from the humiliation of its defeat by Germany in 1870, and the consequent loss of the Rhine provinces of Alsace and Lorraine. A small country with a population of 40 million (scarcely half that of Germany), Britain's industrial position was declining (source 2.6), so colonial markets and manpower were increasingly necessary to its prosperity. In source 2.5, the great South African proconsul, Viscount Milner (1854–1925), outlines in a speech to the Manchester Conservative Club the case for a closer relationship between the mother country and the colonies.

A Russian cartoon drawn in 1904 shows Japan (with sword) being encouraged by her allies to attack Russia's bases in Manchuria. Which countries do you think the characters in the pith helmet and the top hat are meant to represent?

Source 2.4 Lord Kimberley on the need for British intervention in Malaya, 1873

The condition of the Malay Peninsula is becoming very serious. It is the old story of misgovernment of Asiatic States. This might go on without any very serious consequences except the stoppage of trade were it not that European and Chinese capitalists stimulated by the great riches in tin mines which exist in some of the Malay States are suggesting to the native Princes that they should seek the aid of Europeans . . . We are the paramount power in the Peninsula up to the limit of the states tributary to Siam, and looking to the vicinity of India & our whole position in the East I apprehend that it would be a serious matter if any other European Power were to obtain a footing on the Peninsula.

W D McIntyre, p. 205.

Question
- Which 'other European Power' (or Powers) might Kimberley have had in mind?

Source 2.5 Alfred Milner on the importance of 'Greater Britain', 1906

Physical limitations alone forbid that these islands by themselves should retain the same relative importance among the vast empires of the modern world which they held in the days of smaller states—before the growth of Russia and the United States, before united Germany made those giant strides in prosperity and commerce which have been the direct result of the development of her military and naval strength. These islands by themselves cannot always remain a Power of the very first rank. But Greater Britain may remain such a Power, humanly speaking, for ever, and by so remaining, will ensure the safety and the prosperity of all the states composing it, which, again humanly speaking, nothing else can equally ensure. That surely is an object which in its magnitude, in its direct importance to the welfare of many generations, millions upon millions of human beings, is out of all proportion to the ordinary objects of political endeavour. . . .

G Bennett (Ed), p. 352.

Question • In what sense may Milner's words now be seen as predicting later events?

Source 2.6 Britain's position in world industry from 1800 to 1911–13

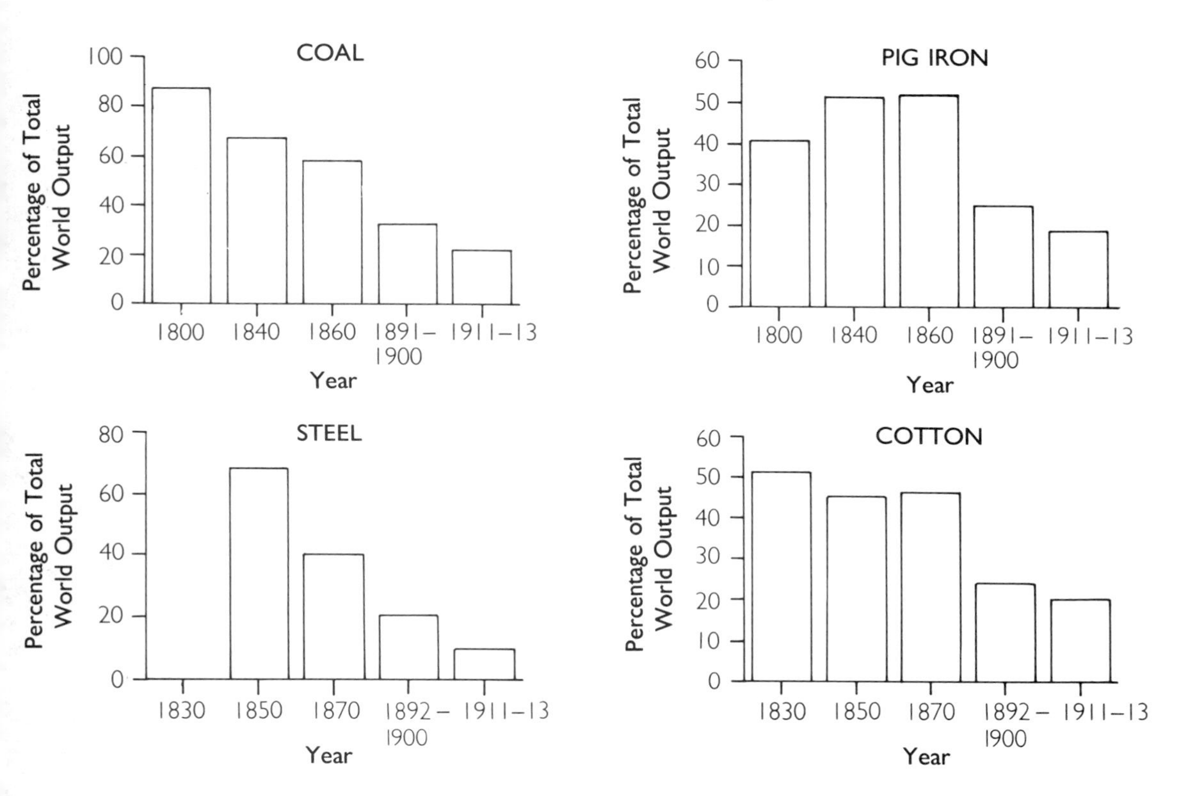

Adapted from E J Hobsbawm, appendix 24.

Question • What were the reasons for this dramatic decrease in Britain's international economic position?

Imperialism of Free Trade

This section examines the evidence for an 'economic' interpretation of imperialism in Asia. In source 2.7, Tate discusses the reasons for the dramatic growth in the industrial demand for tin during the 19th century. This is why Lord Kimberley was worried by the prospect of French or German intervention in Malaya, because Malaya was easily the world's biggest source of the metal.

Source 2.8 is a graph tracing the rise of British foreign investment after 1815. This graph appears to confirm Hobson's contention that **finance capitalism** was an important element of Europe's overseas expansion during the later 19th century. Source

finance capitalism last stage of capitalism, whereby 'surplus' capital is exported wholesale to colonies

2.9 is an excerpt from a speech from the Marquess of Salisbury (1830–1903) to the House of Lords, in which he puts the argument that the Government had a duty to use its power to 'smooth the paths' for British businessmen seeking to break into overseas markets.

However, one might legitimately question whether this is proof of imperialism. Today nations invest overseas and support the commercial activities of their nationals (for example, with trade fairs and high-level negotiations) and few people would call that imperialism. What, if anything, was different about the so-called **free trade imperialism** of the 19th century? One thing, clearly, is that the European governments were not prepared to let Asian countries that did not want or need Western manufactures, like China, isolate and protect themselves from the trading activity around them. In source 2.10, the Chinese Emperor Ch'ien-lung explains to George III of England, in a patronising letter, that his country does not value as precious the same things as Europeans, nor are they in need of Western goods. In 1838, the Chinese Government moved to prevent the profitable (but illegal) smuggling of opium from British India. The British firms involved appealed to their government for compensation, and Prime Minister Lord Palmerston did not let them down, sending an expeditionary force to south China. Three years later, the defeated Chinese signed an **unequal treaty**, the Treaty of Nanking, whose key articles are reproduced as source 2.11.

free trade imperialism theory that period of free trade could be imperialistic although no territory is annexed

unequal treaty treaty imposed forcibly on Asian countries permitting Europeans to live and trade unhindered by local laws

Source 2.7 The growing demand for tin

As a result of the Industrial Revolution a whole new range of uses for tin as an alloy came into existence, particularly in connexion with the rise of the electrical and automobile industries. But the revolutionary increase in the demand for tin after 1800 sprang from the rapid expansion of the tin-plate industry. After 1800 the tin-plate manufacturers were easily the largest single consumers of tin. In 1805 they were already buying up one-third of total world production, by mid-century they were consuming at least half. The industry, which had its origins in fourteenth-century Germany, was now virtually an English monopoly, meeting 'practically all the world's requirements'. The expansion of the tin-plate industry was itself the consequence of the rise of the tin-canning industry. The tin-can, a hermetically-sealed, non-corrosive container for food or drink, was actually invented during the Napoleonic Wars at the beginning of the century, but did not come into its own for another generation. However by the 1850s large urban centres of population in Western Europe had been created by the process of industrialisation, whose inhabitants required the food surpluses being produced in Australia and elsewhere. The tin-plate industry received further stimuli from North America because of the Civil War (1861–5) there, the rise of commercial oil production and the increasing use of corrugated tin roofing. Improved communications facilitated demand, and after the 1840s in particular there was a tremendous upsurge in world tin consumption. Demand reached its peak in the last quarter of the century and doubled again between 1900 and 1930.

D J M Tate, pp. 9–10.

Question

- The rise of the tin industry in Malaya was encouraged by the development of new uses for tin. What demand factors influenced the growth of the Malayan rubber industry?

Source 2.8 British foreign investments, from 1815–20 to 1959

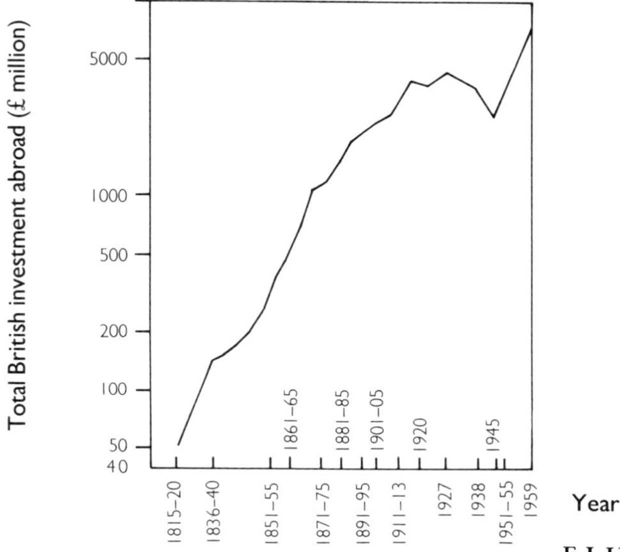

E J Hobsbawm, appendix 32.

Question

- Why did the British make fewer investments abroad during the 1930s?

Source 2.9 Lord Salisbury on the duty of government towards commerce, 1895

Of course the question of administration interests us very much, and we hope that great advantage may be conferred upon the natives by the introduction of English Government, and the enforcement of the peace which accompanies English rule; but the administration of the country is not the sole or the main object that should interest us. It is our business in all these new countries to make smooth the paths for British commerce, British enterprise, the application of British capital, at a time when other paths, other outlets for the commercial energies of our race are being gradually closed by the commercial principles which are gaining more and more adhesion. Everywhere we see the advance of commerce checked by the enormous growth which the doctrines of Protection are obtaining. . . .

***Parliamentary Debates*, 4th series, cs. 698–701.**

Questions

- What is a 'doctrine of protection'?
- Why would the adoption of such a doctrine by Britain's competitors pose a threat to her prosperity?

Source 2.10 The Chinese perspective on international trade, 1793

AN IMPERIAL EDICT TO THE KING OF ENGLAND: You, O King, are so inclined toward our civilisation that you have sent a special envoy across the seas to bring to our Court your memorial of congratulations on the occasion of my birthday and to present your native products as an expression of your thoughtfulness. On perusing your memorial, so simply worded and sincerely conceived, I am impressed by your genuine respectfulness and friendliness and greatly pleased.

As to the request made in your memorial, O King, to send one of your nationals to stay at the Celestial Court to take care of your country's trade with China, this is not in harmony with the state system of our dynasty and will definitely not be permitted. Traditionally people of the European nations who wished to render some service under the Celestial Court have been permitted to come to the capital. But after their arrival they are obliged to wear Chinese court costumes, are placed in a certain residence, and are never allowed to return to their own countries. This is the established rule of the Celestial Dynasty with which presumably you, O King, are familiar. Now you, O King, wish to send one of your nationals to live in the capital, but he is not like the Europeans, who come to Peking as Chinese employees, live there and never return home again, nor can he be allowed to go and come and maintain any correspondence. This is indeed a useless undertaking.

Moreover the territory under the control of the Celestial Court is very large and wide. There are well-established regulations governing tributary envoys from the outer states to Peking, giving them provisions (of food and traveling expenses) by our post-houses and limiting their going and coming. There has never been a precedent for letting them do whatever they like. Now if you, O King, wish to have a representative in Peking, his language will be unintelligible and his dress different from the regulations; there is no place to accommodate him. . .

The Celestial Court has pacified and possessed the territory within the four seas. Its sole aim is to do its utmost to achieve good government and to manage political affairs, attaching no value to strange jewels and precious objects. The various articles presented by you, O King, this time are accepted by my special order to the office in charge of such functions in consideration of the offerings having come from a long distance with sincere good wishes. As a matter of fact, the virtue and prestige of the Celestial Dynasty having spread far and wide, the kings of the myriad nations come by land and sea with all sorts of precious things. Consequently there is nothing we lack, as your principal envoy and others have themselves observed. We have never set much store on strange or ingenious objects, nor do we need any more of your country's manufactures. . .

S-Y Teng and J K Fairbank, p. 19.

Question • What can one infer from this document about Ch'ien-lung's notion of England?

Source 2.11 The Treaty of Nanking, 1842

ARTICLE II.

His Majesty the Emperor of China agrees, that British Subjects, with their families and establishments, shall be allowed to reside, for the purpose of carrying on their Mercantile pursuits, without molestation or restraint at the Cities and Towns of Canton, Amoy, Foochow-fu, Ningpo, and Shanghai . . .

ARTICLE III.

It being obviously necessary and desirable, that British Subjects should have some Port whereat they may careen [keel to one side] and refit their Ships, when required, and keep Stores for that purpose, his Majesty the Emperor of China cedes to Her Majesty the Queen of Great Britain, etc., the Island of Hongkong, to be possessed in perpetuity by Her Britannic Majesty, Her Heirs and Successors, and to be governed by such Laws and Regulations as Her Majesty the Queen of Great Britain, etc., shall see fit to direct.

ARTICLE IV.

The Emperor of China agrees to pay the sum of Six Millions of Dollars as the value of Opium which was delivered up at Canton in the month of March 1839, as a Ransom for the lives of Her Britannic Majesty's Superintendent and Subjects, who had been imprisoned and threatened with death by the Chinese High Officers.

ARTICLE V.

The Government of China having compelled the British Merchants trading at Canton to deal exclusively with certain Chinese Merchants called Hong Merchants (or Cohong) who had been licensed by the Chinese Government for that purpose, the Emperor of China agrees to abolish that practice in future at all Ports where British Merchants may reside, and to permit them to carry on their mercantile transactions with whatever persons they please, and His Imperial Majesty further agrees to pay to the British Government the sum of Three Millions of Dollars, on account of Debts due to British Subjects by some of the said Hong Merchants (or Cohong), who have become insolvent, and who owe very large sums of money to Subjects of Her Britannic Majesty.

ARTICLE VI.

The Government of Her Britannic Majesty having been obliged to send out an Expedition to demand and obtain redress for the violent and unjust Proceedings of the Chinese High Authorities towards Her Britannic Majesty's Officer and Subjects, the Emperor of China agrees to pay the sum of Twelve Millions of Dollars on account of the Expenses incurred. . .

ARTICLE X.

His Majesty the Emperor of China agrees to establish at all the Ports which are by the 2nd Article of this Treaty to be thrown open for the resort of British Merchants, a fair and regular Tariff of Export and Import Customs and other Dues, which Tariff shall be publicly notified and promulgated [proclaimed] for general information. . .

Done at Nanking and Signed and Sealed by the Plenipotentiaries on board Her Britannic Majesty's ship *Cornwallis*, this twenty-ninth day of August, 1842, corresponding with the Chinese date, twenty-fourth day of the seventh month in the twenty-second Year of Taou Kwang.

W L Tung, pp. 428–32.

Question • Why is this treaty seen as 'unequal'?

'he clipper ship *y-ee-moon*, 'built for he opium trade' to China, 1860.

The Civilising Mission

Late 18th century drawing of a Chinese lady sewing; note her tiny feet. Upper class Chinese women customarily had their feet bound at birth, to force them into this unnatural shape.

It is difficult to decide whether the 19th century expansion of territorial empire in Asia was done mainly for selfish or for humanitarian reasons. Certainly, there are many documents of the period which express a desire to assist the 'downtrodden' people of Asia by providing them with a liberal, incorrupt system of government and 'civilising' them according to European standards.

In most cases, however, such statements are made after the event and cannot be assumed to reflect the motives that inspired the original conquest. For instance, the English East India Company had had control of Bengal for more than fifty years when Thomas Babington Macaulay (1800–59) used the Parliamentary debate on the renewal of the Company's Charter to make his impassioned appeal for the adoption of a civilising mission in India (source 2.12). It should be noted, though, that Macaulay couches his appeal partly in self-interested, economic language. And despite the humanitarian arguments for expansion expressed by James Brooke in source 2.13, the same appeal to the profit motive is apparent. This letter was written after Brooke's first visit to Borneo in his yacht *Royalist*, but before he was installed as Governor of Sarawak—becoming, thereby, the first of the White Rajahs—by the embattled Sultan of Brunei.

Source 2.12 Thomas Babington Macaulay on the advantages of civilising India, 1833

To the great trading nation, to the great manufacturing nation, no progress which any portion of the human race can make in knowledge, in taste for the conveniences of life, or in the wealth by which those conveniences are produced, can be matter of indifference. It is scarcely possible to calculate the benefits which we might derive from the diffusion of European civilisation among the vast population of the East. It would be, on the most selfish view of the case, far better for us that the people of India were well governed and independent of us, than ill governed and subject to us—that they were ruled by their own kings, but wearing our broad cloth, and working with our cutlery, than that they were performing their salams to English collectors and English Magistrates, but were too ignorant to value, or too poor to buy, English manufactures. To trade with civilised men is infinitely more profitable than to govern savages.

Parliamentary Debates, **3rd series, cs. 534–6.**

Question • Is a Parliamentary speech generally a good indicator of the views of the politician making it?

Source 2.13 James Brooke on the 'improvement' of Sarawak, 1841

Singapore, March 31.—I have not determined on settling in Borneo without the most mature and serious deliberation, and without seeing a fair prospect of success, and if successful, of no ordinary advancement. It is not only on private views of advantage that I would act, but that I would, generally speaking, seek rather to add to my reputation than my fortune. To develope the resources of a large country is a task I should be most proud to accomplish; and whether we look to the benefits which must accrue to the natives, or to the extension of British trade, it is equally calculated to rouse our best energies. The country of Sarāwak is the finest conceivable; and the influx of Chinese settlers renders its rapid improvement not only possible, but certain, if not impeded by unhappy causes, which it is equally impossible to foresee or calculate upon. Even looking on the undertaking in its worst light, and supposing that, after a year or two, I find it impracticable to accomplish what I so much desire, the attempt will, and must, conduce [contribute] greatly to ameliorate [improve] the native condition, and give them a taste for British manufactures, and some appreciation of a just and protecting government.

Capt R Mundy, pp. 342–3.

Question

- This document is taken from a personal journal (diary). Why would one be inclined to give extra weight to this source?

European Racial Doctrines

de facto existing in fact, whether by right or not; here, where marriage ceremony has not taken place but couple lives as husband and wife

Nordic descended form Germanic peoples of northern Europe

social Darwinism theory that white Anglo-Saxons were entitled to rule over others because they were a superior species

In the 17th and 18th centuries, European men freely associated with Asian women. The Portuguese, particularly, encouraged this in order to populate their settlements and to ease the pain of exile. Although the Dutch and English had no settled policy on intermarriage, they did not forbid it. Consequently, relationships with native women were regarded during this period as quite an acceptable form of social behaviour, and many loving ***de facto*** relationships developed between the races. One such is recorded in source 2.15.

The climate of opinion gradually changed, and by the mid-19th century interracial love-affairs were generally considered cheap and sordid by polite colonial society. What had happened? The English historian, Percival Spear, was a school teacher in India during the 1940s. In source 2.14 he suggests that the prime catalyst for this change was the arrival in the East, from the end of the 18th century, of more European women, which relieved the Companies' servants of the need to seek native companionship.

But other scholars would blame European pseudo-scientists such as the French Comte de Gobineau (1816–82), who wrote *'Essai sur l'inégalité des races humaines'*('Essay on the inequality of human races'). Published in Paris between 1853 and 1855, this was the first statement of the doctrine of **Nordic** superiority and black inferiority, anticipating the theory of **social Darwinism**. In source 2.16, the imperial implications of this doctrine are clearly spelled out by Alfred Milner in an address to the Municipal Congress of Johannesburg in 1903.

'An Afternoon on the Plains' of British India, *c* 1880. European colonials in Asia enjoyed not only plenty of leisure, but the luxury of being waited on by hordes of dutiful servants.

Source 2.14 The widening racial gulf: were women responsible?

One of [the factors responsible for the widening racial gap was] the increasing number of women in the settlements. By another irony the same influence which improved the morals of the settlers increased the widening racial gulf. As women went out in large numbers, they brought with them their insular whims and prejudices, which no official contact with Indians or iron compulsion of loneliness ever tempted them to abandon. Too insular in most cases to interest themselves in alien culture and life for its own sake, they either found society and a house amongst their own people, or in the last resort returned single and disconsolate to Europe. The average Anglo-Indian was equally insular, and his contact had usually first been established by the tyranny of solitude and in time sanctified by custom and tradition. So with the advent of women in large numbers a new standard was introduced, one set of customs and traditions died out, another equally rigid took its place. The attitude of airy disdain and flippant contempt had the background of fear which an unknown and incalculable environment is liable to excite in everyone. For the men the establishment of English homes in place of the prevalent zenanas [part of house reserved for females] withdrew them still more from Indian ways of thought and living, and the acquisition of homes and families gave them something to lose which they had never had before, and thus made them the victims of the same fear. It is this which accounts for the strange panics which from time to time agitate European communities in the east, and for their apparently unaccountable ferocity at time of crisis.

P Spear, pp. 140–1.

Question • Is Spear suggesting that colonial wives were generally more racist than their husbands?

Source 2.15 The will of Lieutenant William Warren, Bengal Army, 1787

In the Name of God Amen I Benjamin William Warren Lieutenant in the service of the Honble East India Company & on the Bengal Establishment do hereby make this my last will & Testament in form & manner following. First I give and bequeath to my natural Daughter, tho' not christen'd yet generally called by me Charlottee Warren, begot upon the Body of my late House Keeper Doordana the sum of Siccon Rupees 10 000 (Ten thousand) to be by my Executors placed at Interest according to the best of their judgment to the greatest advantage & security so as to answer the purpose of clothing her, sending her to England & educating her in a genteel and suitable manner & so as to avoid touching my part of the principal, which is to be her marriage portion, & if at the Age of twenty three years she should not have entered into the marriage state the principal is then to be put into her free & uncontroled possession but prior to that age (excepting only in case of marriage) she is to receive but the Interest accruing from the afore said sum.

S C Ghosh, p. 181.

Question • What can be inferred from this document about the relationship between Lieutenant Warren and Doordana?

Source 2.16 Alfred Milner: Why the white man must rule, 1903

The white man must rule, because he is elevated by many, many steps above the black man; steps which it will take the latter centuries to climb, and which it is quite possible that the vast bulk of the black population may never be able to climb at all. But then, if we justify, what I believe we all hold to, the necessity of the rule of the white man by his superior civilisation, what does that involve? Does it involve an attempt to keep a black man always at the very low level of civilisation at which he is to-day? I believe you will all reject such an idea. One of the strongest arguments why the white man must rule is because that is the only possible means of gradually raising the black man, not to our level of civilisation—which it is doubtful whether he would ever attain—but up to a much higher level than that which he at present occupies. . . .

G Bennett, (Ed), pp. 343–4.

Question • Would Milner have used the same terms about Asians? If not, why not?

The Turbulent Frontiers of Empires

The central paradox of the conquest of Asia by Europeans was, as we have seen, the fact that it was not really planned or intentional. Source 2.17 is taken from a speech of June 1853 on a bill to renew the Charter of the English East India Company for a further 30 years. In it English industrialist and Member of Parliament Richard Cobden (1804–65) restates this paradox and produces a possible resolution: that expansion was the work of the men on the spot in India acting without Parliamentary approval.

If correct, Cobden's analysis begs a further question: *Why* did these local officials act the way they did? Sources 2.18–2.20 consider this problem. Source 2.18, a dispatch from the Governor-General Lord Cornwallis (1738–1805) to his agent at the Maratha capital of Poona, suggests that military action was often, in the first place, defensive; Cornwallis argues in this dispatch that the 'system of neutrality' forced on him by Whitehall (the British Government; see source 2.1) had badly damaged the English East India Company's reputation as a reliable ally.

Another difficulty was that the process of territorial expansion, once begun, tended to acquire its own momentum. Partly this was a function of the lawless conditions that prevailed in many parts of Asia, which posed a threat to the security and economic development of the territories already under European administration. Invariably colonial officials found it cheaper and more effective to enlarge their own frontiers and deal with the trouble directly than to try and enlist, through diplomatic channels, the co-operation of unwilling and frequently powerless Asian rulers. Partly, too, it reflected a very natural disposition on the part of the men on the spot to seek glory in spectacular feats of arms. Victory in a quick and relatively bloodless colonial war, such as that which the United States fought with Spain in 1898, was a guarantee of rapid promotion. Here we highlight the case of American Admiral George Dewey (1837–1917), who became a national hero after sinking eight Spanish warships in Manila Bay (source 2.19).

Finally, while Western governments may not actually have gone out of their way to seek colonies, most were quite happy to accept them if they came their way. After Dewey's victory, the Philippine Islands could not be given back to Spain, and as source 2.20 shows, President McKinley faced a considerable dilemma. The extract is taken from a speech that President McKinley made to a delegation of Methodists at the White House in 1899. (This speech was then written up by James Rusling for an article in the Methodists' newspaper, *The Advocate*, some time later. Rusling may or may not have been a member of the original delegation.) In this speech, McKinley shows that he did not think that the Filipinos were capable of ruling themselves so he decided to keep the islands. It was the beginning of a colonial association that lasted until the end of World War II.

Source 2.17 Richard Cobden on the paradox of continuing annexations in India, 1853

. . . If any good has arisen from our government of the Colonies, it has come from enlightened public opinion, emanating from this country, and chiefly brought to bear on our Colonial Minister in this House. If there be any hope for the amelioration [improvement] of India, it must come from the same source; and I want the Indian Government to have such a tangible, visible form, that the public opinion of this country may be able to reach it, and that there be no mask or screen before it as now. With an enlightened public opinion brought to bear more directly on the affairs of India, there will be a better chance of avoiding . . . [these] constant wars and constant annexation of territory. In other parts of the world, no Minister of the Crown would take credit for offering to annex territory anywhere. On the west coast of Africa, it might not be less profitable to extend our territory than in Burmah; yet a resolution of a Committee of this House, many years ago, forbade the extension of our territories in tropical countries. When an adventurous gentleman, Sir James Brooke, went out and took possession of some territory on the coast of Borneo, the enlightened Government of Sir Robert Peel and his colleagues resolutely resisted all attempts to induce them to occupy any territory there. Recently, when it was announced in this House that orders had been given to the admiral on that station [i.e. Singapore] that on no account should any fresh territory be acquired, the announcement was received with loud cheering. . . .

How is it [then] that this goes on constantly in India, to the loss and dilapidation of its finances? With a declaration in the journals of this House, and in an Act of Parliament never repealed [revoked], that the honour and interest of this country were concerned in not extending its territory in the East, these continual annexations still go on in India. Why do these things happen? It is because at the present time all the authority in these matters is left virtually in the hands of the Governor-General of India.

J Bright and J E T Rogers (Eds), pp. 500–1.

Question • Why was so much authority 'left ... in the hands of the Governor-General of India'?

Source 2.18 Lord Cornwallis' justification for the second Mysore war, 1790

Feb. 28, 1790.

Some considerable advantages have no doubt been experienced by the system of neutrality, which the Legislature required of the Government in this country, but it has at the same time been attended with the unavoidable inconvenience of our being constantly exposed to the necessity of commencing a war, without having previously secured the assistance of efficient Allies.

The late outrageous infraction [violation] of the treaty of peace by Tipu, furnishes a case in point.

We could not suffer the dominions of the Raja of Travancore, who was included by name as our Ally in that treaty, to be ravaged or insulted, without being justly charged with pusillanimity [cowardice] or a flagrant breach of faith, and without dishonouring ourselves by that means in the view of all the powers in India; and as we have been almost daily obliged for several years past, to declare to the Mahrattas and to the Nizam, that we were precluded [prevented] from contracting any new engagements with them for affording them aid against the injustice or ambition of Tipu, I must acknowledge that we cannot claim as a right the performance of those promises which the Mahrattas have repeatedly made to cooperate with us, whenever we should be forced into a war with that Prince.

R Muir (Ed), p. 180.

Question • In declaring war on Mysore, was Cornwallis solely motivated by considerations of 'honour'?

Source 2.19 A poem in praise of Admiral Dewey, 1898

They say that Dewey's coming back
To take a short vacation,
And when he does there'll surely be
A lot of jubilation.
For everybody in the land,
From youngest to the oldest,
Will rush to see the hero who
Is reckoned as the boldest.

They want to see the man who led
His fleet where dangers bristled,
And who was coolest when he stood
Where Spanish missiles whistled;
The man who bravely sailed where Dons
Had big torpedoes scattered,
Who banged away until their ships
To pieces he had battered.

Yes, he's the man they want to see,
And far they'll go to meet him;
They'll strain their eyes as he draws near,
And joyfully they'll greet him.
The women, too, will all turn out,
The matrons and the misses,
And all the pretty girls will try
To favor him with kisses.

Upon him then will be conferred
The freedom of the cities,
And every band in every town
Will play its choicest ditties.
Each orator will hail him with
Most eloquent expressions,
And all the citizens will join
In forming big processions.

Long pent up joy will then break loose,
And like a flood go sweeping,
And on Manila's hero then
All honors we'll be heaping.
Yes, when brave Dewey comes back home
There'll be a grand ovation,
For he's the darling and the pride
Of all this mighty nation.

J L Stickney, p. 367.

Question • Did Dewey's exploits in Manila Bay warrant the frenzied acclamation they received?

Source 2.20 President McKinley explains why he annexed the Philippines, 1899

The truth is I didn't want the Philippines and when they came to us as a gift from the gods, I did not know what to do about them . . . I sought counsel from all sides—Democrats as well as Republicans—but got little help. I thought first we would take only Manila; then Luzon; then other islands, perhaps, also. I walked the floor of the White House night after night until midnight; and I am not ashamed to tell you, gentlemen, that I went down on my knees and prayed Almighty God for light and guidance more than one night.

And one night late it came to me this way—I don't know how it was but it came: (1) that we could not give them back to Spain—that would be cowardly and dishonorable; (2) that we could not turn them over to France or Germany—our commercial rivals in the Orient—that would be bad business and discreditable; (3) that we could not leave them to themselves—they were unfit for self-government—and they would soon have anarchy and misrule over there worse than Spain's was; and (4) that there was nothing left for us to do but to take them all, and to educate the Filipinos, and uplift and civilise and Christianise them, and by God's grace do the very best we could by them, as our fellow-men for whom Christ also died. And then I went to bed, and went to sleep and slept soundly, and next morning I sent for the chief engineer of the War Department (our mapmaker), and I told him to put the Philippines on the map of the United States, and there they are, and there they will stay while I am President!

G J A O'Toole, p. 386.

Questions

- Given that this story was first published in a Christian journal three years after the event, would you say that it was likely to be a faithful record of McKinley's words?
- If not, what words might have been added or altered by Rusling, the *Advocate*'s reporter?

United States cartoon of 1900 showing President McKinley with flag leading the American punishment of the Chinese 'Boxers' for their attacks on Europeans. Where do the words on the flag come from, and what is the significance of the reference to 'treaty rights'?

Role of Public Opinion

vicarious experienced imaginatively, not first-hand

thin red line metaphor for Europeans in colonial frontier situation (British soldiers' uniforms were red)

autocratic where there is absolute government by only one person

By the later 19th century imperialism was not only respectable but immensely popular, satisfying the masses' 'lust' for **vicarious** adventure and public spectacle.[7] One indication of its popularity was the literary success of writers such as G A Henty who wrote approximately 90 tales of heroism on the imperial frontier: these books sold 250 000 copies per year during the 1890s. Another indication was the way the public flocked to events such as the Paris Colonial Exhibition of 1889 and the English Royal Jubilee of 1897. Source 2.21 reveals that the Jubilee was as much a celebration of the British Empire as it was Queen Victoria's 60 years on the throne. A third indication was the success of 'down-market' newspapers like Alfred Harmsworth's *Daily Mail* (1894), which specialised in shocking, sensationalist stories about the **thin red line**. Its notorious, invented account of the 'massacre' of the Europeans besieged in the Peking diplomatic residence by the fanatical Chinese 'Boxers' (source 2.22) is an example.

Nevertheless it cannot be assumed that public opinion caused the New Imperialism of the late 19th century. In Eastern Europe, where governments were still mainly **autocratic**, public opinion did not matter, and evidence shows that British and French governments used public opinion to justify expansion only when they had already decided to act for other reasons.

Source 2.21 The Diamond Jubilee procession, June 1897

Jubilee Day dawned dully, but before long the sun—that sun which never sets on the British Empire—burst forth brightly, and the Queen in her magnificent progress through London brought with her the proverbial 'Queen's weather'. The pageant witnessed in London on Tuesday has never been equalled. Never in the world's history has such an imposing picture of world-wide Empire been displayed. No wonder that the multitudes who witnessed the procession burst into rapturous and enthusiastic cheers as blue-jackets and soldiers, Colonial troops and Indians came by, typifying as they did the Empire and the good swords that won it. The enthusiasm of the spectators reached fever heat when the Gracious Lady in whose honour the celebration was taking place came by. There was something more than the mere expression of loyalty to the head of the British Empire in the cheers and waving of handkerchiefs that greeted the Queen all along the route of the procession. There was besides a feeling of affection and devotion to her personally. Many shed tears as they looked at the silver-haired lady who has ruled this mighty Empire for sixty years, and was being so lovingly greeted by the multitudes of loyal subjects who had assembled to see her triumphal progress through the heart of her Empire to return thanks to Almighty God for the blessings vouchsafed [granted] to her and her people.

The *Graphic*, 26 June 1897, p. 786.

Question • Is this document a 'primary source'? If so, what is it a primary source *for*?

Source 2.22 The Peking massacre that never happened, 1900

The Europeans fought with calm courage to the end against overwhelming hordes of fanatical barbarians thirsting for their blood. While their ammunition lasted they defied Chinese rifle and artillery fire and beat back wave after wave of their assailants. When the last cartridge had gone their hour had come. They met it like men. Standing to their battered defences they stayed the onrush of the Chinese, until, borne down by sheer weight of numbers they perished at their posts. They have died as we would have had them die, fighting to the last for the helpless women and children who were to be butchered over their dead bodies. . . Of the Ladies, it is enough to say that in this awful hour they showed themselves worthy of their husbands. Their agony was long and cruel, but they have borne it nobly, and it is done.

H-M Lo (Ed), pp. 139–40.

Question

- Unlike the previous document, this is a description of an event that never occurred. Is it, nevertheless, a valid primary source?

3 *Pillars of Empire*

. . . the rulers are the merest handful amongst the ruled, a tiny speck of white foam upon the dark and thunderous ocean.

Marquess Curzon of Kedleston, 1904

Having looked at the motives for European expansion in Asia, this chapter examines the methods they employed to attain their objectives. Few in numbers, the Europeans nevertheless succeeded in carving out for themselves vast empires—in some cases much bigger and with more people than the countries they came from. How did they do it? How did so few rule over so many for so long?

KEY DATES

1746 Madras successfully defended by 900 French-trained troops against a 10 000-strong native army under the command of the Nawab of the Carnatic.

1823 The *Diana*, the first steam-powered gunboat to operate in Asian waters, launched at Kidderpore Dock, Bengal.

1825 Pangaran (Prince) Dipanagaro raises the standard of revolt in eastern Java. The ensuing Java War (1825–30) costs 200 000 Indonesian lives.

1842 A British expeditionary force to Afghanistan of 4500 men was almost wiped out while retreating from Kabul to Jalalabad.

1846 (Sir) James Brooke becomes the first White Rajah (ruler) of Sarawak

1847 Vietnamese Emperor Thieu-tri (*r*. 1841–48) decrees that all Europeans found in his domains will be executed.

1851 Beginning of the 66 year reign of Charles Brooke (b. 1829), the second White Rajah of Sarawak.

1857 Outbreak of the 'Great Rebellion' in India; 200 European women and children massacred at Cawnpore (Kanpur).

1867 The Snider-Enfield breechloading rifle introduced into the British Army.

1877 Spectacular Darbar (pageant) held in Delhi to celebrate Queen Victoria's assumption of the title of 'Empress of India'.

1882 Overcoming fierce local resistance, French troops capture the citadel of Hanoi, Tongking (Vietnam).

1884 First reliable machine-gun (capable of firing 11 rounds per second) developed by American inventor (Sir) Hiram Maxim (1840–1916).

1908 The Dewa Agung of Klungkung leads his unarmed followers in a suicidal *puputan* (final battle) against Dutch troops at Badoeng, Bali.

1919 General Reginald Dyer (1864–1927) shoots and kills 379 Indian demonstraters at Jallianwallah Bagh, Amritsar.

ISSUES AND PROBLEMS

So far we have looked at the motives which lay behind European expansion in Asia—the 'why' of imperialism. This chapter shall examine the means of expansion—the 'how'. The aim will be to identify the factors which enabled the Europeans to impose, and maintain their military and political control over vast tracts of Asia for more than 200 years.

With hindsight, it is tempting to regard the process of European global expansion and empire-building as inevitable, as something predestined or, as the imperialists themselves would have put it, ordained by God. However, far from being inevitable, the conquest of Asia can fairly be regarded as having occurred against the odds. Europe was remote: until the opening of the Suez Canal in 1869 it was over 8000 kilometres (or up to six months) away by sea. In the later 18th century, when the English and Dutch Companies made their first major conquests, the West was not widely industrialised and did not possess a significant technological advantage in weapons over the larger non-Western States. As shown in chapter 2, expansion was not a high priority of the home authorities. What is more, the conquest, pacification and colonial exploitation of Asia were carried out by a relatively tiny number of Europeans and Americans. At its peak, in the 1930s, the white colonial workforce numbered about half a million—only 0.05 per cent of the native population. Only a small fraction of these actually served in the administration. Thus the highly regarded Indian Civil Service (ICS) was staffed by less than 1000 British residents, the North Borneo Civil Service by just 45. The Brookes' Kingdom of Sarawak, as a French visitor discovered to his amazement, was run by some 'thirty Englishmen . . . [with the aid of] a few hundred native soldiers and policemen'.[1]

subjugation act of bringing under the control of

However, it should not be imagined that the Asians were easily defeated militarily. The **subjugation** of Asia took time. As early as 1580, the Philippine Islands were effectively under the dominion of Spain; and by the 1750s Java and Bengal were virtually under Dutch and British occupation. However, the British did not make much progress in south India until 1799, in northern India until 1803, and in western India until 1818. Awadh was annexed in only 1856—nearly 100 years after the Battle of Plassey which was the beginning of empire-building in Bengal. Similarly, it took the Dutch nearly a century to consolidate their control over the outer islands of Indonesia, the pacification of Bali being completed only in 1908 (source 3.22). Territorially, the Western empires in Asia attained their maximum extent in 1922—barely two decades before the Japanese invasion of South-East Asia brought the **hegemony** of the white man to an abrupt halt. Moreover, the majority of Europe's conquests in Asia were bought at high cost. Nearly 8000 East India Company troops died in the Burma War of 1824–6, and another 5000 died in the Afghanistan campaign of 1842. After the Battle of Ferozepur, the opening engagement in the first Anglo–Sikh War of 1845, the Governor-General, Lord Hardinge commented, 'One more such victory and we are done'.[2] Everywhere in Asia the Europeans met fierce resistance.

hegemony political overlordship; moral/cultural domination by one group or country over another

Nevertheless, the Europeans *did* prevail, and this success needs to be explained.

THE DEBATE

The obvious answer is that the Europeans won out because they were militarily superior; and this thesis has been persuasively put by a number of scholars.

> From the very beginning, the basis of European imperialism was superior force: From the gunned ship of the sixteenth century to the gunned helicopter of today, advanced military technology has supported this expansive drive.[3]

Others, however, would strongly contest this claim that military technology was a significant factor in the initial period of expansion. V G Scammel, for example, noting that early model guns were notoriously unreliable in the damp climate of the tropics, concludes that 'Neither in Asia nor anywhere else was empire the inevitable consequence of technological superiority'.[4]

Alternatively, it has been suggested that the long life of colonial rule in countries like India had much to do with the personality and performance of the officials who ran the administration. For instance, the ICS has been likened by Philip Mason—a former member—to the ideal governing élite described by Plato in his *Republic*.[5]

caste hereditary Hindu class whose members follow a common occupation, share food and intermarry

divide-and-rule strategy to gain power by splitting one's opponents

A third theory holds that the Europeans triumphed because the opposition—fragmented by **caste**, race and religion—was easily divided. 'India', writes Sir John Seeley, 'had no jealousy of the foreigner, because India had no sense whatever of national unity. . . . there *was* no India and therefore, properly speaking, no foreigner'.[6] This is the **divide-and-rule** theory.

Finally, a number of historians have recently referred to the fact that, although European colonial rule was vigorously resisted by some Asians, it was warmly welcomed by others. According to this view, colonialism ultimately rested on the implied consent of the masses and on the active collaboration of a significant section of the native élite.

The Thin Red Line

With the significant exception of Siberia, which absorbed some nine million Russian emigrants between 1800 and 1914, Asia did not appeal to Europeans as a place for settlement. Over most of the vast continent the climate was extremely harsh and disease was rife in the tropics. Moreover all the fertile agricultural regions of Asia were already densely settled by peasant farmers. Unlike the nomadic hunters and gatherers of the American prairies or the South African veldt, these entrenched agriculturalists were not removed easily from their lands. Thus, in contrast to the Americas and Australasia, colonisation played little part in Europe's imperial design.

When English Parliamentarian Charles Dilke (1843–1911) visited Ceylon in 1866 during a whirlwind tour of the Empire, he was amazed by how few white people there were (source 3.1). Even by the 1930s, things had not changed greatly (source 3.2). Indeed in some of the more jungly parts of Asia—like Borneo—the European population was likely to consist entirely of a couple of administrators and missionaries and their wives. With legitimate pride, A B Ward, who served as a district officer in Sarawak under the second 'White Rajah' Charles Brooke, recalls in his memoirs that he and two English subordinates together controlled a region of 6000 square miles (source 3.3). How did this tiny élite manage to impose its authority on a huge continent inhabited by hundreds of millions of alien and often warlike peoples?

Chandni Chowk, the main bazaar in Old Delhi, 1885. How many Europeans can you find in this picture?

Source 3.1 Charles Dilke on tour in Ceylon, 1866

The first thing that strikes the English traveller in Ceylon is the apparent slightness of our hold upon the country. In my journey from Galle to Columbo, by early morning and mid-day, I met no white man; from Columbo to Kandy, I travelled with one, but met none; at Kandy,

I saw no whites; at Nuwara Ellia, not half-a-dozen. On my return, I saw no whites between Nuwara Ellia and Ambe Pusse, where there was a white man in the railway-station; and on my return by evening from Columbo to Galle, in all the thronging crowds along the roads there was not a single European. There are hundreds of Cinghalese in the interior who live and die, and never see a white man. Out of the two and a quarter millions of people who dwell in what the planters call the 'colony of Ceylon,' there are but 3000 Europeans, of whom 1500 are our soldiers, and 250 our civilians. Of the European non-official class, there are but 1300 persons, or about 500 grown-up men. The proposition of the Planters' Association is that we should confide the despotic government over two and a quarter millions of Buddhist, Mohamedan, and Hindoo labourers to these 500 English Christian employers.

C W Dilke, pp. 417–8.

Question • Would you say that Dilke was sympathetic to the 'proposition of the Planter's Association'?

Source 3.2 Europeans in Asian colonies as a proportion of total population, *c.* 1930

	Philippines	*India*	*Indonesia*	*Vietnam*	*Malaya*
otal population	14 m	350 m	61 m	19 m	7 m
uropeans	9000	96 000	300 000	40 000	24 000
uropeans to total population	1 : 1550	1 : 3650	1 : 200	1 : 475	1 : 290

R Jeffrey (Ed), p. 5.

Question • How would you account for the somewhat larger proportion of Europeans living in Indonesia?

Source 3.3 District administration in Sarawak under the Brookes

The 2nd Division, *my* district, I felt entitled to call it now, comprised four good-sized rivers—the Batang Lupar, Saribas, Kalaka and Sebuyau. It covered six or seven thousand square miles, with a population of about 60 000 Dyaks, as well as a considerable number of Malays and Chinese. Three English Officers controlled the whole of this territory.

A B Ward, p. 32.

Question • How is a 'Dyak' different from a 'Malay'?

Impact of Modern Technology

It is understandable that European statesmen and colonial administrators were reluctant to emphasise the military basis of their rule. They preferred to stress the good relationship which allegedly existed between the natives and themselves; over the entire period of European hegemony in Asia, force was not used often as a means of control.

coerce force into obedience, persuade using force

Nevertheless the colonial regimes did not lack the means to **coerce** the natives. They all maintained large standing armies and police forces, and, as Sir James Fitzjames Stephen (the same who talked to the boys at Eton) points out in source 3.4, 60 000 English troops were not stationed in India for fun. When danger threatened, the authorities did not hesitate to call out the troops, and the instinctive reaction of the Europeans on the spot was always to reach for a gun. In source 3.5, highschool teacher Philip Richards provides an eyewitness account of the Punjab 'troubles' of 1919 in letters to his mother in England. His version of events is exaggerated, but the letters give a marvellous insight into Richards' state of mind as he contemplated the approach of the 'mob'.

Furthermore, by the 19th century, the European forces were clearly superior to their native opponents in regard to their weapons. This gap widened more and more as time went by and as Western military hardware improved rapidly. In source 3.6, the American scholar D R Headrick reminds us that these advances in military technology were characteristic of the period, and the natives were unable to keep up with them. Better armed and generally better led, European armies gained an impressive record of victories over much more numerous Asian forces from the mid-18th century onwards. Indeed, defeats became so rare that when, in 1842, a numerically superior British army, equipped with artillery, was all but wiped out in Afghanistan, the Prime Minister, the Duke of Wellington (1769–1852), upon hearing the dire news, automatically attributed it to cowardice. As one of those responsible for the downfall of Tipu Sultan in the Mysore War of 1799, the Prime Minister should have known what he was talking about. The truth is that superior technology did not always compensate for over-extended supply lines, harsh terrain, and fanatical native resistance.

Source 3.4 Foundations of the government of India, 1883

It is not improbable that in the course of time, though I think it will be a long time, native habits of life and ways of thought will give way to, and be superseded by, those of Europe. Should that happen, the bulk of the population might come not merely to submit to European rule, but in some degree to like it, and to sympathise with its spirit. What changes in the system of government this might involve no one can say. Till, if ever, that time arrives, it will never in my opinion be safe for the British Government to forget for a moment that it is founded not on consent but on conquest; that it must, if it exists at all, be absolute, because its great and characteristic task is that of imposing on India ways of life and modes of thought which the population regards, to say the least, without sympathy, and to which it might easily be brought to feel active dislike, though they are essential to its permanent wellbeing and to the credit of its rulers. There is a practical proof of the truth of what I have said, which appears to me unanswerable. It is the fact that we maintain in India an army one-third of which consists, or ought to consist, of sixty thousand British troops, amongst whom are comprised the whole of the artillery. What are they there for? Obviously to sustain the British power. Would that power be maintained if they were permanently withdrawn? I do not believe that any one in this country upon whom the slightest responsibility for his words rests, or can ever rest, will answer this question in the affirmative. But if the maintenance of a great army, one-third of which consists of British troops, while the other two-thirds are officered by Englishmen, is the indispensable condition of British rule in India, who will say that the power is not essentially belligerent? or deny that, as long as it is to exist at all, it must be absolute, in the sense of not being controlled by a representative assembly or assemblies?

J F Stephen, pp. 558–9.

Question • Why were Stephen's remarks considered rather provocative?

Source 3.5 Two letters from Lahore, 1919

14 April 1919

The hartal [closing of businesses as protest] was on, as the Sirdar told us; and there was every prospect of a violent mob overrunning the European quarter. This was on Thursday last. Norah and I were invited to join forces with our nearest European neighbours—Post Office people, whose friendship we had recently made. We went over to their house, and while Norah talked to the lady and her little girl, the Deputy Postmaster-General for the Punjab, Baluchistan, and Aden, asked me to step into his bedroom. *The first thing*—on his bed were three rifles, a revolver, and a display of ammunition. *The second thing*—I was not to tell the ladies—five Europeans murdered in Amritsar (30 miles off); the banks burned; station partly set on fire; etc. Hence, he advised us to sleep in their house for the night. Norah thought it unnecessary, and may have been surprised by my firmness in deciding to send for our camp-beds. We got across with our things just before dark; and then, after dark, heard the mob howling. A weird sound—an angry mob; separated from you by a thin line of men in khaki.

23 April 1919

. . . Was it last Thursday it all began or was it the Thursday before? It seems a long time! The authorities in Lahore had had news of the murders in Amritsar when the mob in Lahore began to march up the Mall with the design of getting into the civil station—where Europeans preponderate. If they had got in, the scenes at Amritsar would probably have been re-enacted upon a larger scale. They were stopped by a handful of police who fired their rifles—and a few cavalry. Representatives of the mob are now saying that they only intended a harmless march and that there was no need for the police to fire. No officer, however, would have been justified in allowing the mob to proceed under any circumstances; still less after hearing the news—which many in the mob must also have known—of the murdered Europeans in Amritsar.

No, Lahore has had a lucky escape from attempted massacre.

P E Richards, pp. 180–1, 184.

Question • Why are letters generally considered a good source for writing history?

Source 3.6 The evolution of the rifle

At the beginning of the nineteenth century the standard weapon of the European infantryman was the muzzle-loading smoothbore musket with a bayonet. The Brown Bess, which British soldiers used up to 1852, was much the same weapon their forefathers had used at Blenheim in 1704. This gun had an official range of 200 yards, but even at half that distance it was so inaccurate that soldiers were advised to withhold their fire until they saw the whites of their enemies' eyes. Even so, said the gunmaker W W Greener, they commonly shot away their weight in lead for every enemy they killed. Since muzzle-loaders took a minute or more to reload, they were more useful as pikes than as guns. . .

Another important advance was the percussion cap. Until the early nineteenth century, gunpowder had been ignited with a flintlock, a method which worked only in dry weather. In 1807 Alexander Forsyth introduced the use of fulminates as priming powders, and in 1816 Thomas Shaw patented the copper percussion cap. In tests by the Woolwich Board of the British army the new Brunswick percussion-cap rifle misfired only 4.5 times per thousand rounds, as compared with 411 times per thousand for flintlocks. As a result of these tests a few select British units were equipped in 1836 with Brunswick rifles. The impact of these guns can be judged from this account of a battle near Canton in 1841: 'A company of sepoys, armed with flintlock muskets, which would not go off in a heavy rain, were surrounded by some thousand Chinese, and were in eminent peril when two companies of marines, armed with percussion-cap muskets, were ordered up, and soon dispersed the enemy with great loss.'

The third important advance was the cylindro-conoidal bullet, developed to overcome the inaccuracy of the muzzle-loader. Ideally a bullet should be small enough to slip down the barrel easily, yet large enough to grip the rifling on the way out. Early efforts concentrated on making the bullet swell at the moment of firing. Of these, the most successful was that of Minié, whose bullet was long and pointed, with a plug at the back to make it expand. Not only did the Minié bullet take the rifling and spin well, but its streamlined shape helped give it a flat trajectory. The results were amazing. At 100 yards the Minié rifle hit the target 94.5 per cent of the time, compared to 74.5 per cent for the Brunswick; at 400 yards the figures were 52.5 and 4.5 per cent respectively . . . This stage in the evolution of the gun reached its peak in 1852–53 when the British army replaced the Brown Bess with the Enfield rifle which fired the new bullets. This was the first European military gun to be made on the 'American system' of interchangeable parts. Its great advantage, like that of the French Minié, was its accuracy; it had an official range of 1200 yards and an effective one of 500 yards, five or six times greater than the Brown Bess. . .

The last bit of 'progress' in the evolution of the gun arose in response to the special needs of empire. In the words of the historians of guns, Ommundsen and Robinson; '. . . savage tribes, with whom we were always conducting wars, refused to be sufficiently impressed by the Mark II bullet; in fact, they often ignored it altogether, and, having been hit in four or five places, came on to unpleasantly close quarters.' The solution to this unpleasantness was patented in 1897 by one Captain Bertie-Clay of the Indian ammunition works at Dum-Dum: the mushrooming or 'dum-dum' bullet. This particular invention was so vicious, for it tore great holes in the flesh, that Europeans thought it too cruel to inflict upon one another, and used it only against Asians and Africans.

By the 1890s the gun revolution was complete. Most European infantrymen could now fire fifteen rounds of ammunition in as many seconds, lying down undetected, in any weather, with an effective range of up to half a mile. Machine gunners had even more power. Though the generals were not to realise it for many decades, the age of raw courage and cold steel had ended, and the era of arms races and industrial slaughter had begun.

D R Headrick, pp. 249–50, 253–4, 256.

Question
- What is Headrick's point of view towards **a** imperialism and **b** technology?

he 3rd Punjab
olunteer Rifle
:ompany, Sukkur,
larch 1883.

Power of Personal Authority

In general, though, the Europeans did not have to rely on the firepower of their armies and police forces to keep order; the subject populations living under colonial rule mostly remained quiet and submissive. Why? One factor was the personal authority or **charisma** exercised by the best colonial administrators. Colonials might be stereotyped as wooden autocrats, but most, at least in the 19th and 20th centuries, were well educated, hard working and courageous, and respected by those whom they governed. (Sir) Stamford Raffles (1781–1826), for example, became a revered figure during his time as Lieutenant-Governor of Java (1811–16). According to his confidential assistant, natives wept when he left (source 3.7a). Years afterwards he was described in god-like terms to an American traveller, Walter M Gibson, by Saiyyid Sherif Ali, the Arab headman of Palembang (source 3.7b). Similarly, John Nicholson, who died during the recapture of Delhi in 1857, was idolised by the tough, rustic tribesmen of Hazara on India's northwest frontier (source 3.8).

charisma quality of personal magnetism enabling rare individuals to sway emotions of others

What is more, the Europeans who served in Asia were brought up to think of themselves as racially destined to rule, and this made them supremely confident. Something of this spirit, which allowed English women to confront angry mobs without flinching, can be seen in source 3.9, which comes from an address to the Royal Colonial Institute by Mrs Douglas Cator, who had spent some time in Malaya.

Source 3.7 Sir Stamford Raffles: Two native tributes, 1816 and 1850

a Mr Raffles was accompanied to the bank by all the respectable inhabitants of Batavia who took leave of him with tears in their eyes. The chief Chinese and Native inhabitants would not take leave of him till they had seen him on board, when they evinced the deepest grief on taking leave. All his intimate friends came off on board with him and here I am not capable of describing the distressing scene which took place.

M Collis, p. 86.

b The Panyorang spoke of Raffles, the Tuan Besar Ingres,—the English Great man, as Sir Stamford Raffles, the famous British Governor in the Archipelago, and enlightened founder of Singapore, is called and remembered by the Malays of Palembang. He said that the people did not believe that the great good man was dead, and looked for his coming again.

W M Gibson, p. 128.

Question • Why might a biographer of Raffles pay more heed to the second extract than the first?

Source 3.8 The 'Nikalsaini' cult in the Punjab, 1860

When they heard of his glorious death, they came together to lament, and one of them stood forth and said there was no gain from living in a world that no longer held Nikalsain. So he cut his throat deliberately, and died. Another stepped forward and said that was not the way to serve their great *Guru*; that if they ever hoped to see him again in a future state, and to please him whilst they lived, they must learn to worship Nicholson's God. The rest applauded, and off started several of them, and coming to Peshâwar, went straight up to the missionary there, and told him their desire.

Capt L J Trotter, pp. 314–15.

Question • What does this document reveal about the process of Christian conversion in Asia?

Source 3.9 Why I am proud of being an Englishwoman, 1909

The more I see of the Colonies, the more I see of the world, the prouder I am of being an Englishwoman. Our national characteristics of justice and honour and pluck and our sense of fair play have given us a power of colonisation, a success where others fail, and a position in every quarter of the globe which no other nation can touch. Nothing perhaps is more touching in our tropical Colonies than the way the natives trust in us and in our judgment. We are the only pucka [genuine] white nation to the Malay, and nothing to their minds is beyond our power, from protecting them single-handed against their enemies to healing them of every disease, including paralysis.

One year, when we were up in the interior of Borneo, we found the river tribes very nervous after two cowardly murders, one of the murderers being still at large. A whole settlement moved down to where we were, so as to be under the magic shadow of our wing—two unarmed people, one of them only a woman; but we were English, and that, in their minds, was everything. You meet exactly the same spirit among the African tribes. They are all just like children in their absolute confidence in us, and great is our responsibility when we abuse their faith, which is just what, unfortunately, we do at times. . . .

P H Kratoska (Ed), p. 294.

Question • 'In colonialism it was appearances that mattered, not reality.' How does the above document bear out that assertion?

Myth of European Invincibility

European authority in Asia was based partly on actual power, partly on arrogance and bluff, and partly on the willingness of Asians to believe that the Europeans really were all that they boasted they were—super-human beings. When he started work as a district officer in Punjab early in the 20th century, (Sir) Malcolm Darling found that Indians belonging to the peasant class held white people, whom they rarely saw, in awe (source 3.10). Sita Ram Pandey's recollections of his first day as a *sepoy* (soldier) in the Indian Army (source 3.12) help us to understand why.

While illiterate peasants were possibly more receptive to myths about white 'gods' than their more worldly city cousins, they were not alone in their conviction that Europeans were a superior species. Even a well-educated Asian bureaucrat like Feng Kuei-fen (1809–74), who interacted frequently with foreigners in the Chinese 'treaty-ports' in the course of his official duties during the 1840s and 1850s, could think of no other adequate explanation to their startling success (source 3.11).

:aving for Simla, 1912.

These myths took root and spread, and they became self-fulfilling, as the Europeans strove to act the part that their subjects expected them to play. Leonard Woolf recalls that in Ceylon, where he served in the early 1900s, the British unconsciously sought to mould themselves into replicas of characters in stories by Rudyard Kipling.[7] Similarly, Sir Walter Lawrence, who served as Lord Curzon's Private Secretary during his period as Viceroy (1898–1905), talks in his memoirs of the 'mutual make believe' which underlay the mystique of the British Raj (source 3.13).

Source 3.10 The aura of the topee, *c.* 1905

In those far-off days the British Raj inspired such awe in the peasant that you had only to wear a topee to be treated like royalty. One evening I had been out for a ride and two water-carts, each drawn by a yoke of oxen, were jogging along ahead of me when their drivers, suddenly spotting me, leapt from their carts and rushed them, oxen and all, into the ditch, one to the left, the other to the right, so that I could have the road to myself as I passed between them. This kind of thing happened again and again. If, too, a horseman met you on the road—'and they are legion'—he would either alight until you had passed, or turn into a neighbouring field to avoid having to do this. No wonder that this excessive respect for authority went in some cases to the head.

M Darling, p. 21.

Question • Do you think that, as a young man, Darling enjoyed being 'treated like royalty' in India?

Source 3.11 A Chinese Mandarin speculates on the reasons for his country's weakness, 1861

The most unparalleled anger which has ever existed since the creation of heaven and earth is exciting all who are conscious in their minds and have spirit in their blood; their hats are raised by their hair standing on end. This is because the largest country on the globe today, with a vast area of 10,000 *li*, is yet controlled by small barbarians. . . According to a general geography by an Englishman, the territory of our China is eight times larger than that of Russia, ten times that of America, one hundred times that of France, and two hundred times that of England. . . Yet now we are shamefully humiliated by those four nations in the recent treaties—not because our climate, soil, or resources are inferior to theirs, but because our people are really inferior. . . Why are they small and yet strong?

S-Y Teng and J K Fairbank, p. 52.

Question • What was wrong with Feng's logic?

Source 3.12 Sita Ram Pandey's first meeting with an Englishman, 1814

After bathing, and eating the morning meal, my uncle put on full regimentals and went to pay his respects to the Adjutant *sahib*, and Commanding Officer. He took me with him. I was rather dreading this because I had never yet seen a *sahib* and imagined they were terrible to look on and of great stature—at least seven feet tall! In those days there were only a few *sahibs* in Oudh; only one or two *sahib* Residents in Lucknow, where I had never been. In the villages of my country the most extraordinary ideas existed about them, and any one who had chanced to see a *sahib* told the most curious stories. In fact nothing was too farfetched to be believed. It was said that they were born from an egg which grew on a tree, and this idea still exists in remote villages. Had a *memsahib* come suddenly into some of our villages, she would, if young and handsome, have been considered to be some kind of fairy, and would probably have been worshipped; but should she have been old and ugly, the whole village would have run away to hide in the jungle, believing her to be a witch. It is therefore hardly surprising that I should have been so terrified at the prospect of seeing a *sahib* for the first time in my life.

I remember once, when I was attending a fair at the Taj Mahal in Agra, an old woman said she had always believed that *sahibs* came from eggs which grew on a tree; but that morning she had seen a *sahib* with a fairy by his side. The fairy was covered with feathers of the most beautiful colours, her face was as white as milk, and the *sahib* had to keep his hand on her shoulders to prevent her from flying away. All this the old woman had seen with her own eyes, and she swore it was true. I am not so ignorant now, of course, but I would have believed it when first I arrived at Agra. I afterwards often saw that *sahib* driving out with his lady. She wore a tippet made from peacock feathers, and the old women had mistaken this for wings.

We went to the Adjutant's house, which was four times the size of the headman's house in my village. He was on the verandah, with a long stick, measuring young men who were recruits. He was very young, not as tall as myself, and had no whiskers nor moustache. His face was quite smooth and looked more like a woman's than a man's. This was the first *sahib* I had ever seen, and he did not fill me with much awe. I did not believe he could be much of a warrior with a face as smooth as that since among us it is considered a disgrace to be clean-shaven; in fact a smooth-faced soldier is usually the butt for many jokes.

J Lunt (Ed), pp. 12–13.

Question

- Sita Ram's memoirs were written in Urdu around 1863 and translated into English by an Army officer in 1873. How does this affect their value as a historical source on the early 19th century?

Source 3.13 The mystique of the 'heaven-born', 1928

Our life in India, our very work more or less, rests on illusion. I had the illusion, wherever I was, that I was infallible and invulnerable in my dealing with Indians. How else could I have dealt with angry mobs, with cholera-stricken masses, and with processions of religious fanatics? It was not conceit, Heaven knows: it was not the prestige of the British Raj, but it was the illusion which is in the very air of India. They expressed something of the idea when they called us the 'Heaven born,' and the idea is really make believe—mutual make believe. They, the millions, made us believe we had a divine mission. We made them believe they were right. Unconsciously perhaps, I may have had at the back of my mind that there was a British Battalion and a Battery of Artillery at the Cantonment near Ajmere; but I never thought of this, and I do not think that many of the primitive and simple Mers had ever heard of or seen English soldiers. But they saw the head of the Queen-Empress on the rupee, and worshipped it. They had a vague conception of the Raj, which they looked on as a power, omnipotent, all-pervading, benevolent for the most part but capricious, a deity of many shapes and many moods.

W Lawrence, pp. 42–3.

Questions

- How does this view of the Raj compare with that presented in source 3.4?
- Which do you think is closer to the truth?

European Terrorism of the Native Peoples

conquistadors
Spanish conquerors

The pre-modern period in Asia was not notable for its gentleness, but even by the harsh standards of the time, the Portuguese and Spanish ***conquistadors*** became a byword for cruelty and ruthlessness. In source 3.14, American historian Daniel Boorstin gives some examples of their inhumane behaviour. However, as Boorstin observes, the sadism of the Portuguese was not without reason: it had a military purpose, namely to terrorise their enemies into surrender.

terrorism calculated use of extreme violence as political/psychological weapon

Terrorism was also used by later European colonial regimes to deter rebellion, and in particular to revenge acts of violence against whites. In India, in May 1857, a mutiny broke out in the ranks of the *sepoy* army and quickly spread to sections of the civil population. Many Europeans were murdered, including some 200 women and children whose bodies were unceremoniously thrown down a well at Cawnpore (Kanpur). When a British 'flying column' retook the city in September, its commander, Brigadier-General J G S Neill, tried to come up with a punishment befitting this terrible crime. The resulting regimental order is graphically described for us by an eyewitness, Sergeant William Forbes-Mitchell of the 93rd Sutherland Highlanders (source 3.15).

Bodies of Indian rebels killed by Sir Colin Campbell's Highlanders litter the courtyard of the Secundra Bagh, Lucknow.

Although the British in India never descended again to this level of barbarity, they came close, on a smaller scale. In the Punjab in 1919, Brigadier-General Reginald Dyer, the Chief Martial Law Administrator of Amritsar, ordered his troops to fire on a political meeting in a local park known as Jallianwallah Bagh. With 379 people dead and hundreds more wounded, the Jallianwallah Bagh shooting is justly referred to as a massacre. At the time, however, Dyer was hailed by the Tories in England as a hero who had 'saved' the Raj, and he himself mounted a vigorous defence of his actions before a commission of enquiry (source 3.16).

The British were not unique in their use of terror tactics against colonial subjects. Source 3.17 is a contemporary account of the repressive measures taken by the French after the Yen Bay uprising in Vietnam in 1930. Ironically, though, overkill of this sort usually served merely to stiffen native resistance to colonialism. Thus, in the long term, terrorism backfired against its perpetrators.

Source 3.14 Portuguese atrocities in the 16th century

With ships heavily armed for battle, they were uninhibited in using terror. We have seen how Vasco da Gama cut up the bodies of casually captured traders and fishermen, and sent a basketful of their hands, feet and heads to the Samuri of Calicut simply to persuade him into a quick surrender. Once in power, the Portuguese governed their India in the same spirit. When Viceroy Almeida was suspicious of a messenger who came under a safe-conduct to see him, he tore out the messenger's eyes. Viceroy Albuquerque subdued the peoples along the Arabian coast by cutting off the noses of their women and the hands of their men. Portuguese ships sailing into remote harbors for the first time would display the corpses of recent captives hanging from the yardarms to show that they meant business.

D Boorstin, p. 192.

Questions

- Did the Portuguese use excessive cruelty?
- What excuses (if any) could one offer for their conduct?

Source 3.15 The vengeance of Brigadier-General Neill, 1857

After trial and condemnation, all prisoners found guilty of having taken part in the murder of the European women and children, were to be taken into the slaughter-house by Major Bruce's *mêhter* police, and there made to crouch down, and with their mouths lick clean a square foot of the blood-soaked floor before being taken to the gallows and hanged. This order was carried out in my presence as regards the three wretches who were hanged that morning. The dried blood on the floor was first moistened with water, and the lash of the warder was applied till the wretches kneeled down and cleaned their square foot of flooring. This order remained in force till the arrival of Sir Colin Campbell in Cawnpore on the 3rd of November, 1857, when he promptly put a stop to it as unworthy of the English name and a Christian Government.

W Forbes-Mitchell, p. 20.

Question

- Why did Neill force his prisoners to lick the floor before hanging them? (Note: most of the captured *sepoys* were Brahmins.)

Source 3.16 General Dyer's evidence before the Hunter Commission of Enquiry into the Punjab 'disturbances', 1919

Q When you heard of the contemplated meeting at 12-40 you made up your mind that if the meeting was going to be held you would go and fire?

A When I heard that they were coming and collecting I did not at first believe that they were coming, but if they were coming to defy my authority, and really to meet after all I had done that morning, I had made up my mind that I would fire immediately in order to save the military situation. The time had come now when we should delay no longer. If I had delayed any longer I was liable for court-martial.

Q Supposing the passage was sufficient to allow the armoured cars to go in would you have opened fire with the machine-guns?

A I think, probably, yes.

Q In that case the casualties would have been very much higher?

A Yes.

Q And you did not open fire with the machine-guns simply by the accident of the armoured cars not being able to get in?

A I have answered you. I have said if they had been there the probability is that I would have opened fire with them.

Q With the machine-guns straight?

A With the machine-guns.

Q I gather generally from what you put in your report that your idea in taking this action was really to strike terror? That is what you say. It was no longer a question of dispersing the crowd but one of producing a sufficient moral effect.

A If they disobeyed my orders it showed that there was complete defiance of law, that there was something much more serious behind it than I imagined, that therefore these were rebels, and I must not treat them with gloves on. They had come to fight if they defied me, and I was going to give them a lesson.

Q I take it that your idea in taking that action was to strike terror?

A Call it what you like. I was going to punish them. My idea from the military point of view was to make a wide impression.

Q To strike terror not only in the city of Amritsar, but throughout the Punjab?

A Yes, throughout the Punjab. I wanted to reduce their *morale*; the *morale* of the rebels.

Lord Hunter, pp. 188–90.

Question • What does this exchange reveal about **a** the 'military mind' and **b** General Dyer's state of health in 1919?

Source 3.17 French repression in Vietnam, 1930–33

The hysterical repression that followed these native demonstrations led to great cruelties. Mass arrests took place, and torture, according to press reports, was widely applied, especially at Thu-Duc, Saigon and Cholon, where electrical torture was used. According to Andree Viollis . . . the methods also included: Deprivation of food, bastinado [caning on soles of feet], pins driven under the nails, half-hanging, deprivation of water, pincers on the temples, (to force the eyes outward) and a number of others that are not printable. One may be quoted: 'With a razor blade, to cut the skin of the legs in long furrows, to fill the wound with cotton and then burn the cotton.'. . .

The infantry, chiefly of the Foreign Legion and the Colonial Legion, was even more destructive [than the air attack]. The accounts of the brutalities are far from pleasant reading, nor is atrocity-mongering a useful form of propaganda. But since some of the Legionaries were charged with an excess of zeal later at a trial at Hanoi in June, 1933, it is possible to extract certain portions of the record that will show on whom lies the responsibility for this behaviour. The extracts are taken from La Franche-Indochine and deal with the hearing of June 12:

. . . Lieutenant Lemoanne: I received the orders of Commandant Lambert to kill all prisoners. On occasion I captured Communists *in flagrante delicto* [red-handed] and executed them forthwith.

The President: You had prisoners tortured.

Lamoanne: That was to influence the population. . . .

Captain Doucin: Precise instructions were given in confidential message 280 of 8/10/30, ordering the execution of every communist caught *in flagrante delicto*. . .I am well aware that bloody deeds were done. But who were shot? Communists! Well, I don't think enough were shot. That's my opinion. . . .

The President [to Legionnaire Palowski]: Had you received instructions to execute prisoners?

Palowski: Yes, instructions from M. Robin, who afterwards congratulated us and said: *Tres bien! Continuez*!

The accused were acquitted.

R Postgate, pp. 44–5.

Question • Is this a 'primary' or a 'secondary' source?

Absence of Sustained Native Resistance

We have already referred in several places to the resistance offered to colonialism by Asian rulers and peoples. Clearly, resistance was widespread and, at times, quite formidable. The Indian 'mutiny' of 1857 brought about the collapse of British power over much of central north India, and in Awadh (Oudh) particularly, the Raj was only restored after the local chiefs had been bought off with promises of land and tax-relief (see source 5.7). Likewise, the Dutch hold on Java hung in the balance for nearly three years following the rebellion of Pangeran (Prince) Dipanagara, a son of the Sultan of Jogjakarta, in 1825. Yet, in both cases, the Europeans prevailed in the end.

As the century wore on, while no less willing, Asian resistance became increasingly fruitless and desperate. For example, by 1908, the people of Bali in eastern Indonesia knew they had lost the fight. Yet, as Vicki Baum recounts in her fictionalised history of the island (source 3.18), their pride did not allow surrender. Dressed in ceremonial white robes, and wielding their *krisses* (daggers) the people of Klungkung village, led by their headman, launched a suicidal charge against the massed firepower of General Veldt's modern army. They were brutally mown down. By the 1930s, when the English novelist Somerset Maugham was writing, Malaya at least had become a haven of tranquillity (source 3.19).

Source 3.18 The 'Battle' of Badoeng, Bali, 1908

It was then that people were seen emerging on to the top of the steep steps leading down from the gateway of the puri to the road below. They were clothed in white and adorned with flowers. Their tread was sedate and they had lances in their hands. Behind them were bearers carrying a man shoulder-high on a decorated throne. They set the throne down. For a moment the man was quite alone on the top-most step. Then a procession of other men came out one after another and ranged themselves behind the solitary figure. Dekker let go of his sword-hilt. He could not understand what it all meant. It looked like a scene on the stage. Yes, that was it—it was exactly like a scene at the opera in Amsterdam, to which he had been occasionally . . .

Suddenly the lord [that is, the Raja of Badoeng] raised his head and in a flash he had drawn the kris from the scabbard that projected behind his shoulders. He held it outstretched above his head and it flashed in the sun. An unearthly shout broke from the men around him. The next moment they charged the Dutch, kris in hand.

'Fire!' the captain shouted. 'Fire!' Dekker roared in his unpractised voice of command. The men fired. A few of the Balinese fell and lay where they had fallen. The rest charged up the main road towards the turning from Tian Siap where the howitzers were firing and the bugles blared. The two companies of the 11th followed at the double. The whole scene was wrapped in the smoke and dust that hung like a curtain between them and the wild rush of the white-clad figures and almost hid it from sight. Dekker ran in front of his men between the shell-shot walls on either side of the road. A party of men armed with spears dashed out of one of the doorways. Dekker was surprised to find that he had drawn his short sword. A bayonet flashed past him into the stomach of a Balinese. When he looked about him he found that he was separated from his company. It looked as though the Dutch would give way before the mad assault of the Balinese. Dekker shouted a command. He saw one of his men fall and noted the astonishment in his face. The other officers, too, were shouting to their men and they re-formed their ranks. And now they had joined up with the main body.

Gun- and rifle-fire swept the Balinese as they came round the Tian Siap turning and charged straight for the Dutch troops. The lord was the first to fall. The rest ran on over his dead body in a wild onset and when they fell, still more came on. A mountain of wounded and dead was piled up between the puri and the Dutch troops. Meanwhile the gate-way disgorged more and more of them, all with krisses in their hands, all with the same death-frenzy in their eyes, all decked out and crowned with gold and flowers.

Three times the Dutch ceased fire, as though to wake these frantic people from their trance or to spare and save them. But the Balinese were set on death. Nothing in the world could have arrested them in their death-race, neither the howitzers nor the unerring aim of the sharp-shooters, nor the sudden stillness when the firing ceased. Hundreds fell to the enemy's rifles, hundreds more raised their krisses high and plunged them into their breasts, plunging them in above the collar-bone so that the point should reach the heart in the ancient, holy way . . .

Dekker, for one, was unable to endure the sight of men killing their wives and then themselves, and of mothers driving a kris into their infants' breasts. He turned away and vomited.

V Baum, pp. 500–2.

Question • This extract is taken from a 'historical novel' about the Bali uprising. How does it differ from a conventional historical account?

Source 3.19 Somerset Maugham reflects on the British presence in Malaya

The countries of which I wrote were then at peace. It may be that some of those peoples, Malays, Dyaks, Chinese, were restive under the British rule, but there was no outward sign of it. The British gave them justice, provided them with hospitals and schools, and encouraged their industries. There was no more crime than anywhere else. An unarmed man could wander through the length of the Federated Malay States in perfect safety. The only real trouble was the low price of rubber.

W S Maugham, pp. 7–8.

Question • Why might this be considered a *nostalgic* view of Malaya in the 1930s?

Why Resistance Failed

heterogeneity being composed of different elements

Why, then, did it take so long (until the mid-20th century) for resisters to dislodge the Europeans from their occupation of south and South-East Asia? One reason was clearly the military strength of the Europeans, which made them virtually impregnable to conventional attack, even by larger forces. Indeed many a brave Asian commander, like Vietnam's Hoang Dieu (1829–82), the Viceroy of Hue, was driven to despair by their inability to 'resist the enemy' (source 3.20).

Another reason was the **heterogeneity** of Asian society which made unified resistance difficult. This enabled the Europeans to play cynical games of 'divide-and-rule' with their opponents.

However, both of these arguments are based on the assumption that Asians wanted to be rid of the Europeans. This was not necessarily the case, as the final two excerpts show. In fact, in the 19th century, European rule seems, if anything, to have been rather popular. A plaque was commissioned in 1833 to commemorate the reforming Governor-Generalship of Lord William Bentinck (1774–1839). Although the words on the plaque were composed by T B Macaulay, who had come out to India as the first Law Member of the Governor-General's Council, the monument itself (source 3.21) was planned and paid for by Indians. And when Keshub Chandra Sen (1838–84) spoke in Calcutta in 1877 on the theme of loyalty to the Queen, he was loudly cheered (source 3.22). In the end, colonial rule survived in Asia because resisters were outnumbered by collaborators (see chapter 5).

Villing collaborators: ckshaw coolies ply their rade in Simla, summer apital of the Indian mpire.

Source 3.20 The suicide note of Viceroy Hoang Dieu, 1908

In the second month of this year, many French battleships sailed toward northern Vietnam and most of them moored not far from Hanoi. The population of the citadel became uneasy at their approach.

I ventured to think of Hanoi as a pass opening to all the regions of northern Vietnam. It is certainly a strategic stronghold. Were this place to fall into enemy hands, the rest of the territory would sooner or later follow suit.

For this reason, I both sent instructions secretly to the governors of the neighboring provinces to warn them and memorialised the throne to ask for reinforcement. Several times, however, Imperial edicts rebuked me for excessive concern with military matters or for not knowing the proper way to resist the enemy.

I searched myself and found that I was not in command of any real power. Nevertheless I could not with decency forsake my duties as I am, after all, an important official. Conforming my conduct to that of my predecessors, I maintained an unwavering loyalty to the King . . .

While our plans and preparations for defense were still indefinite, the enemies suddenly broke their earlier agreement.

On the seventh day [24 April 1882] they submitted their ultimatum and on the following day they unleashed the main force of their troops against the citadel.

The enemy troops surrounded us, numerous as ants.

The Western cannon exploded, deafening as thunder.

In the city, fire spread over all the houses in every street.

In the citadel, fear wrung the heart of the whole population.

Although I had just recovered from an illness, I made every effort to command our troops. We killed more than one hundred enemy soldiers. We succeeded in defending the citadel for longer than half a day.

But what else could we have done? . . .

B L Truong, pp. 110–11.

Question • What excuses does Hoang Dieu offer for losing Hanoi?

Source 3.21 Inscription on the Bentinck Memorial, Calcutta (erected 1833)

TO
WILLIAM CAVENDISH BENTINCK,
who during seven years ruled India with eminent prudence,
integrity, and benevolence;
who, placed at the head of a great empire, never laid aside the
simplicity and moderation of a private citizen;
who infused into oriental despotism the spirit of British freedom;
who never forgot that the end of government is the happiness
of the governed;
who abolished cruel rites;
who effaced humiliating distinctions;
who gave liberty to the expression of public opinion;
whose constant study it was to elevate the intellectual and moral
character of the nations committed to his charge;
THIS MONUMENT
was erected by men
who, differing in race, in manners, in language, and in religion,
cherish with equal veneration and gratitude
the memory of his wise, upright, and paternal administration.

P W Wilson (Ed), p. 455.

Question

- The Memorial was commissioned and paid for largely by Indians, but the words were composed by an Englishman, T B Macaulay. Can it be regarded as an expression of Indian opinion?

Source 3.22 Keshub Chandra Sen on the providential nature of British rule, 1

Were you present at the magnificent spectacle at Delhi, on the day of the assumption of the imperial title by our sovereign? Some men have complained that no religious ceremony was observed on the occasion, and indeed opinion is divided on this point. None, however, can gainsay the fact that the whole affair, from beginning to end was a most solemn religious ceremony, and I rejoice I am privileged to say this in the presence of our noble-hearted Viceroy. Was any devout believer in Providence there? To him I appeal. Let him say whether the imperial assemblage was not a spectacle of deep moral and religious significance. Did not the eye of the faithful believer see that God Himself stretched His right hand and placed the Empress' crown upon Victoria's head? [Loud cheers.] And did he not hear the Lord God say unto her: 'Rule thy subjects with justice and truth and mercy, according to the light given unto thee and thy advisers, and let righteousness and peace and prosperity dwell in the Empire'? [Applause.]

Would you characterise this sight and this sound as a visionary dream? Is there no truth in the picture? Who can deny that Victoria is an instrument in the hands of Providence to elevate this degraded country in the scale of nations, and that in her hands the solemn trust has lately been most solemnly reposed? Glory then to Empress Victoria! [Applause.]

W T de Bary (Ed), *Indian Tradition*, pp. 618–19.

Question

- How is the force of this document enhanced by the reporter's interpolations (in square brackets)?

4 Colonial Policy and Practice

We must at present do our best to form a class who may be interpreters between us and the millions whom we govern; a class of persons, Indian in blood and colour, but English in taste, in opinions, in morals, and in intellect.

Thomas Babington Macaulay, 1833

THIS CHAPTER LOOKS AT what the Europeans hoped to achieve by taking over large tracts of Asian territory. These hopes and aspirations are measured against the administrative and social record of major colonial regimes in south and South-East Asia during the later 19th and early 20th centuries. This period was the heyday of what came to be known as the 'civilising mission'.

KEY DATES

1813 Christian missionaries allowed into British India for the first time.
1829 The Indian custom of *sati* (widow-burning) outlawed by British Governor-General Lord William Cavendish-Bentinck (1774–1839).
1830 'Culture' (that is, forced cultivation) System introduced into Java by Dutch Governor-General Johannes van den Bosch (1780–1864).
1835 English East India Company abandons its 'Orientalist' policy in favour of a Westernising one, resolving, in future, to spend money only on English-language medium education.
1839 Russian alphabet imposed on the Muslim peoples of Central Asia.
1854 The Indian Civil Service becomes the first public service in the Western world to recruit by competitive examination.
1860 Anglo-French force occupying China sacks and burns the beautiful Imperial Summer Palace near Peking (Beijing).
1891 Age of consent in India to sexual intercourse within marriage fixed by law at 12 years.
1900 Dutch adopt an assimilationist (see below) 'Ethical Policy' in the Dutch East Indies.

ISSUES AND PROBLEMS

One of the major justifications offered for expansion in Asia was that European management was necessary to 'rescue' its 'benighted' peoples from the consequences of 'heathen' superstition and dynastic oppression. From a humanitarian—especially Christian—viewpoint, Asian society was riddled with 'barbarous and cruel customs'[1] which cried out for reform. The reformers also thought that Asians were entitled to share the fruits of Europe's expanding technology. In addition it was considered that the same liberal institutions that had been established over centuries of struggle in the West should be available to Asians. These institutions included elected parliaments, a free press, and an independent judiciary. The first French Governor of Cochin China, Admiral Le Grandiere (1863–68), wrote that the country was a 'land . . . open to civilization . . . and to the fertile ideas of Europe'.[2] And similar hopes were held for India. 'I feel you will agree with me', wrote Governor-General Lord Mayo (1869–71) in a letter to his superior, the Secretary of State in London, '. . . that we have no right to be here at all unless we use all our power for the good of the Blacks'.[3]

rhetoric language used to impress or persuade

To what extent, though, was this imperial **rhetoric** translated into administrative action? Words are cheap; and the statements of people in powerful positions should always be taken with a grain of salt. Were these expressions of humanitarian intent sincere? Or were they mere propaganda, designed to appease liberals and clergymen back home? There are at least two good reasons why we should be sceptical about their sincerity. First, as chapter 3 showed, the European colonial rulers were capable, when cornered, of acts of extreme barbarity. Second, if Lord Macaulay's speech to Parliament in 1833 (source 2.12) can be taken as typical, many European statesmen advocated a policy of taking 'civilisation' to Asia at least partly on economic grounds. By definition, a 'civilised' Asian was more likely to appreciate (and buy) Western manufactures. This much seems clear. Whether for humanitarian or commercial reasons, European colonial administrators in Asia felt that their interests would be served by promoting significant structural change. This led them to take an **interventionist** approach to government in contrast to the previous native leaders. Tax revenue, enhanced by more efficient means of collection, was used to finance schemes of 'public works', health-care and education. Legislation, which could be enforced by the courts, was used to regulate public behaviour. To some extent the success of colonialism can be measured by what it achieved in terms of these administrative yardsticks. For example, during the later 19th century the British Raj oversaw in India the development of the world's largest canal irrigation system and its third largest railway network. Similarly, the American regime (which took control in the Philippines in 1900) must be applauded for its public health measures, which caused the number of cholera cases to decrease from about 100 000 to 820 in fifteen years. Overall, the colonial period was a time of rapid social and economic change.

interventionist favouring government intervention in otherwise non-government affairs

Nevertheless, we should be cautious about claiming a simple cause and effect relationship for this process. The Europeans were generally reluctant to take the initiative for fear of stirring up trouble, especially in relation to religious issues. The result was that many important pieces of social legislation (for example, the 1891 Age of Consent Bill in India) were only proceeded with after strong representations from educated native opinion. As source 4.19 shows, some of the most far-reaching and disruptive changes which occurred under colonialism were brought about by Westernising policies which backfired. In the same way, we should be wary about accepting gross statistical indicators of material growth as evidence of social improvement. Railways clearly did make life easier, but they also assisted foreign goods to penetrate Asian markets and colonial troops to put down native revolts. Similarly, while European-owned plantations contributed significantly to the exports of countries like Malaya and Ceylon, very little of the resulting profits found their way back into the local community.

THE DEBATE

There are three main schools of thought about colonial policy in Asia. The first holds that the 'civilising mission' was nothing more than an elaborate deception, and that the colonial regimes were interested only in development for purposes of enrichment. Many scholars from former colonial dependencies agree with this view. Thus, the distinguished Bengali Marxist historian, Sumit Sarkar, opens his study of modern India with this sarcastic comment:

> The illusion of permanence held powerful sway over the minds of the British in India in 1885, eight years after the Empire had been proclaimed at a grandiose Durbar held in the midst of famine. An ideology of **paternalistic** benevolence, occasionally combined with talk of trusteeship and training towards self-government, thinly veiled the realities of a Raj uncompromisingly white and despotic.[4]

paternalistic
tendency of colonial regimes to treat subjects like children

The second school, by contrast, is prepared to take the colonialists at their word. Whatever their failings in practice, say this group, the colonial regimes did have benevolent intentions, and most assuredly did not set out to exploit their subjects. 'The myth of imperial profit-making', insists David Fieldhouse, 'is false'.[5]

Finally, a number of writers have argued that colonial policy in Asia was actually quite conservative. Far from wishing to remake Asia in their own image, the Europeans preferred, by and large, to tread a softly-softly line designed to satisfy vested interests and minimise the chances of unrest.

Goals and Aspirations of European Colonials

Lord William Cavendish-Bentinck, Governor-General of India 1828–35.

What did the European colonial regimes do with the lands that they had fought so hard to acquire? To answer this question we must first establish what was on the Europeans' agenda, that is, how they planned to reform and develop their newly won possessions. For example, perhaps the two most influential Indian Governors-General of the 19th century were Lord William Bentinck and Lord Dalhousie (1812–60). Like most British colonial administrators of that time, Bentinck and Dalhousie were both interventionists, committed to the utilitarian philosophy of using the resources of government to produce the greatest good for the greatest number. Both men were optimistic about what could be achieved by this means. In source 4.1, taken from his minute on the Hindu rite of *sati* (widow-burning), Bentinck explains why he has resolved to legislate to purge India of this barbaric custom. In source 4.2, Dalhousie writes enthusiastically about the economic and social benefits that India will obtain through the construction of railways.

social engineering structuring of society according to a government's preconceived plan

Ethical Policy policy of Dutch administration in Indonesia giving local officials more power and putting more money into agriculture and education

Significantly, though, this interventionist phase came to an abrupt end in 1857 with the outbreak of the Indian mutiny. By the end of the 19th century optimism about Asia's growth had been replaced by a mood of bleak resignation, exemplified in Kipling's enigmatic poem, '*The White Man's Burden*' (source 4.3).

Indeed, considering colonial Asia as a whole, it would be fair to say that **social engineering** was not a major priority. In the Dutch East Indies, an **Ethical Policy** was belatedly introduced after 1900, but its social effects had only begun to be felt when the policy was abandoned two decades later as being too dangerous politically.

Source 4.1 Lord William Bentinck's minute on the suppression of *sati*, 1829

To consent to the consignment year after year of hundreds of innocent victims to a cruel and untimely end when the power exists of preventing it is a predicament which no conscience can contemplate without horror . . . [Indeed], I should be guilty of little short of the crime of multiplied murder if I [were to] hesitate in the performance of this solemn obligation. I have been already stung with this feeling. Every day's delay adds a victim to the dreadful list, which might perhaps have been prevented by a more early submission of the present question. . .

The first and primary object of my heart is the benefit of the Hindus. I know nothing so important to the improvement of their future condition as the establishment of a purer morality, whatever their belief, and a more just conception of the will of God. The first step to this better understanding will be dissociation of religious belief and practice from blood and murder. They will then, when no longer under this brutalising excitement, view with more calmness acknowledged truths. They will see that there can be no inconsistency in the ways of Providence, that to the command received as divine by all races of men, 'no innocent blood shall be spilt,' there can be no exception; and when they shall have been convinced of the error of this first and most criminal of their customs, may it not be hoped that others, which stand in the way of their improvement, may likewise pass away, and that thus emancipated from those chains and shackles upon their minds and actions, they may no longer continue, as they have done, the slaves of every foreign conqueror, but that they may assume their first places among the great families of mankind? I disown in these remarks, or in this measure, any view whatever to conversion to our own faith. I write and feel as a legislator for the Hindus, and as I believe many enlightened Hindus think and feel.

R Muir (Ed), pp. 293–6.

Questions

- What, in this context, is a 'minute'?
- Could Bentinck fairly be described as a 'cultural imperialist'?

Source 4.2 Lord Dalhousie: Minute on railways, 1853

Great tracts are teeming with produce they cannot dispose of. Others are scantily bearing what they would carry in abundance, if only it could be conveyed whither it is needed. England is calling aloud for the cotton which India does already produce in some degree, and would produce sufficient in quality and plentiful in quantity if only there were provided the fitting means of conveyance for it from distant plains to the several ports adapted for its shipment. Every increase of facilities for trade has been attended, as we have seen, with an increased demand for articles of European produce in the most distant markets of India; and we have yet to learn the extent of value of the interchange which may be established with people beyond our present frontier, and which is yearly and rapidly increasing.

Ships from every part of the world crowd our ports in search of produce which we have, or could obtain in the interior, but which at present we cannot profitably fetch to them; and new markets are opening to us on this side of the globe, under circumstances which defy the foresight of the wisest to estimate their probable value or calculate their future extent.

It needs but little reflection on such facts to lead us to the conclusion that the establishment of a system of railways in India, judiciously selected and formed, would surely and rapidly give rise within this empire to the same encouragement of enterprise, the same multiplication of produce, the same discovery of latent source, to the same increase of national wealth and to same similar progress in social improvement; that have marked the introduction of improved and extended communication in various Kingdoms of the Western world.

O S Nock, pp. 14–15.

Question • What kind of 'social improvement' could Dalhousie have envisaged as resulting from the building of railways?

Source 4.3 The White Man's Burden (*The United States and the Philippine Islands*), 1899

Take up the White Man's burden—
  Send forth the best ye breed—
Go bind your sons to exile
  To serve your captives' need;
To wait in heavy harness
  On fluttered folk and wild—
  Your new-caught, sullen peoples,
  Half devil and half child.

Take up the White Man's Burden—
  In patience to abide,
To veil the threat of terror
  And check the show of pride;
By open speech and simple,
  An hundred times made plain,
To seek another's profit,
  And work another's gain.

Take up the White Man's burden—
  The savage wars of peace—
Fill full the mouth of Famine
  And bid the sickness cease;
And when your goal is nearest
  The end for others sought,
Watch Sloth and heathen Folly
  Bring all your hope to nought.

Take up the White Man's burden—
  No tawdry rule of kings,
But toil of serf and sweeper—
  The tale of common things.
The ports ye shall not enter,
  The roads ye shall not tread,
Go make them with your living,
  And mark them with your dead! . . .

R Kipling, pp. 323–4.

Question • What does Kipling mean by the expression 'White Man's burden'?

Effectiveness of Colonial Administration

It was one thing to want to change Asia according to some preconceived design, but it was quite another to put such schemes into practice. Thinly scattered, and vastly outnumbered by their millions of subjects, the Europeans could do little personally except give orders to their Asian subordinates and hope that they were carried out efficiently.

They were also handicapped by a lack of detailed knowledge of the societies that they were governing. This state of affairs continued until quite late in the history of colonialism when some reliable census data started to become available.

A third obstacle to putting their plans into practice was that the Europeans faced tremendous passive resistance to change among the more traditional sections of the population, such as the peasantry. John Beames (1837–1902) served as a district officer in many parts of eastern India in the period immediately after the Mutiny. Source 4.4, taken from his frank and often tongue-in-cheek memoirs, implies that the much-praised Indian Civil Service was prone to **nepotism** and filled with many rather mediocre performers. From the memoirs of Malcolm Darling (see source 3.10), source 4.5 illustrates the extent to which inexperienced British officers depended on the advice of their native subordinates.

nepotism
appointment of relatives to jobs, often undeservingly

In Source 4.6, another Indian administrator, Thomas Stoker, the land-revenue Settlement Officer for Bulandshahr District, explains to his superiors why he has been unable to win the confidence or gain the co-operation of the villagers in his charge.

Source 4.4 John Beames on the reason for his appointment as Collector and Magistrate of Purnea District, 1862

The new work and new surroundings of my position in Purneah caused me at first some embarrassment. To be appointed to officiate as Magistrate and Collector of a district after only four years' service was unprecedentedly rapid promotion. Even if I had spent those four years in Bengal, becoming familiar with the Bengal system, the charge of so large a district would have been a difficult task for so young an officer. But having passed nearly all the time in the Punjab, where the system was entirely different, it was still more difficult for me to do the work. Purneah was a district of more than 5000 square miles, and was officially described about this time as 'the most lawless district in Bengal'. Owing to its notorious unhealthiness it was shunned by all officers whose ability or personal influence were sufficient to secure them favourite districts. Thus it was generally entrusted to men of inferior calibre, or to men who were in disgrace, for the Government has the habit of reserving some of the most unhealthy districts as 'penal settlements' to which they send any man whom they wish to punish. I was sent there because I was a stranger whom no one knew or cared about. Peacock, the son of Sir Barnes Peacock, Chief Justice of Bengal, who was of the same standing as myself and was appointed Collector at the same time, got the healthy and favourite district of Monghyr. But malarious Purneah was good enough for a friendless man like me.

J Beames, p. 133.

Questions

- What did 'collectors' collect?
- Does this document suggest that British administration in Purnea was efficient?

Source 4.5 Sir Malcolm Darling's first case, *c.* 1905

With the New Year real work began. I now had my own Court and with it my first case. A letter describes the scene. 'I sit on a dais screened by a railing from the touch of the vulgar. On my table repose vast tomes of law, into which from time to time I peer wisely. My Munshi [secretary], who fortunately speaks English, without which justice would be at a standstill, sits on my right. I am as a babe in his arms. It is fearful to think of the power he wields.' My chaprassi [sergeant-at-arms], a stout fellow of six feet two, stands in the body of the Court, and with eyes fixed dog-like upon my face, he tries to penetrate my every thought that an order may be carried out almost before it is given. 'Now the complainant enters, with hands folded beseechingly on his breast. My first duty is to take down his statement. This he makes in Punjabi, of which I know not one word, but the Munshi does, God bless him.' He translates it into Urdu. Even so, I am not much the wiser, but gradually it begins to dawn on me that the man has quarrelled with a neighbour about a rope taken from a well which they own in common. The neighbour, he says, gave him 'three heavy blows'. 'Any marks?' I ask. Whereupon 'he strips his upper half (not very much to come off) and reveals a great tawny trunk with two little spots like pin pricks'. Applying a wise provision of the law, I dismissed the complaint as frivolous.

M Darling, p. 16.

Question

- Was justice done in this case?

Source 4.6 Thomas Stoker on his relations with the villagers of Bulandshahr District, 1887

My daily work for five months in the year brings me into contact with the people in their fields and villages; I am surrounded by them for hours every day; for weeks together I speak to nobody else; I see them under every condition, and hear all their complaints. It is part of my business to visit every one of their homesteads, and to note generally their style of living, their appearance, the character of their houses, their surroundings, their stock and equipment. It is impossible that any person of ordinary intelligence and observation can mix on these terms with the people and remain in ignorance of such broad facts as to whether they are sufficiently fed and clothed and properly equipped for the business of agriculture. If no more than this were required, I could at once and with some confidence give an answer to the inquiry which is now made. . . . But . . . generalizations are not required, . . . an inquiry is to be instituted into specific cases. This I feel bound to represent that I do not think I can carry to a useful or safe conclusion. It is the object of every person who lives by the land to place the condition of himself and his industry before the Settlement Officer in the most disparaging light. It will be useless for me to say that the inquiry has no connexion with settlement; I will not be believed. I cannot divest myself of my official character. Every man whom I question will believe I am seeking a basis on which to assess his rent or revenue, and he will answer accordingly. He will declare that his fields do not return even the seed and labour, and that he and his family are starving. The evidence of my own sight will show him to be lying; but unless I make an inquisition and hunt up evidence, the record will misrepresent the facts. And, indeed, the evidence being that of his fellows, will most likely support than contradict him. These are not mere speculations. I find . . . that since I have been engaged in settlement work my relations with the people are much changed. I am regarded as an enemy, to be opposed by their only weapon, that is to say, deception. I tried at first when going among the fields and villages to help the people by explaining to them such matters as the great improvements which have recently been made in the methods of well-sinking, or the better methods of cultivation I had seen followed, or better staples grown in other parts of the country. I desisted only when I found myself credited with the Machiavellian policy of seeking in this manner fresh grounds and reasons for raising the revenue. My object in making a house-to-house inquiry will certainly be misunderstood, and the facts will certainly be misrepresented. I can answer for the accuracy of my own observations; but I do not think the information supplied by cultivators or proprietors will be equally trustworthy.

E Whitcombe, pp. 268–9.

Question

- How do you think Stoker's superiors would have responded to this report?

Signs of Progress

Nevertheless, despite all the obstacles to change described in the previous section, 'progress' (as the Westerners liked to think of it) did occur. For example, in British India during the 19th century there was a massive expansion in the amount of area under agriculture as banditry was reduced and irrigation extended, especially in the arid north-west.

Likewise, the progressive imposition of British rule over Burma between 1825 and 1885 led to a spectacular increase in the cultivation of rice, which soon became that country's staple export (source 4.9). Jules Harmand writes that the French annexation of Cochin China resulted in a taxation bonanza (source 4.7).

Indians watch with a mixture of fear and wonder as a steam traction engine passes through their village.

Figures on productivity should be interpreted cautiously. In addition to the possibility that the figures could be wrong (we know that governments have before used figures deliberately to flatter their performance), it must be remembered that higher taxation can result from other things than a rise in output (harsher collection measures, for example); gross figures on acreage can mask glaring inequalities in land-ownership; and rising land prices (source 4.8), while indicative of a healthy market, are not necessarily proof of rural prosperity. The 'developmental' record of the colonial regimes was, however, generally better than their social welfare record. Even an imperialist such as Lord Curzon was driven to question the Raj's performance in education (source 4.10).

Source 4.7 Agricultural progress in Cochin China, 1910

The Mekong delta, with fertility above average and ordinarily free of severe climatic problems (though, it is true, with insufficient population to exploit it properly), could furnish the court at Hué before our arrival no more than two millon francs in direct taxes, paid in cereals and in zinc sapeks. After a few years, this same area and this same population provided the French administration with that amount tenfold in silver coin, even though peace was still troubled and before the land could be improved in ways rendered both costly and difficult by the nature of the country.

P D Curtin (Ed), p. 301.

Question • What factors, besides 'fertility', could have produced a tenfold increase in the revenue collected from the Mekong delta?

Source 4.8 Land values in the Madras Presidency, from 1823–24 to 1878

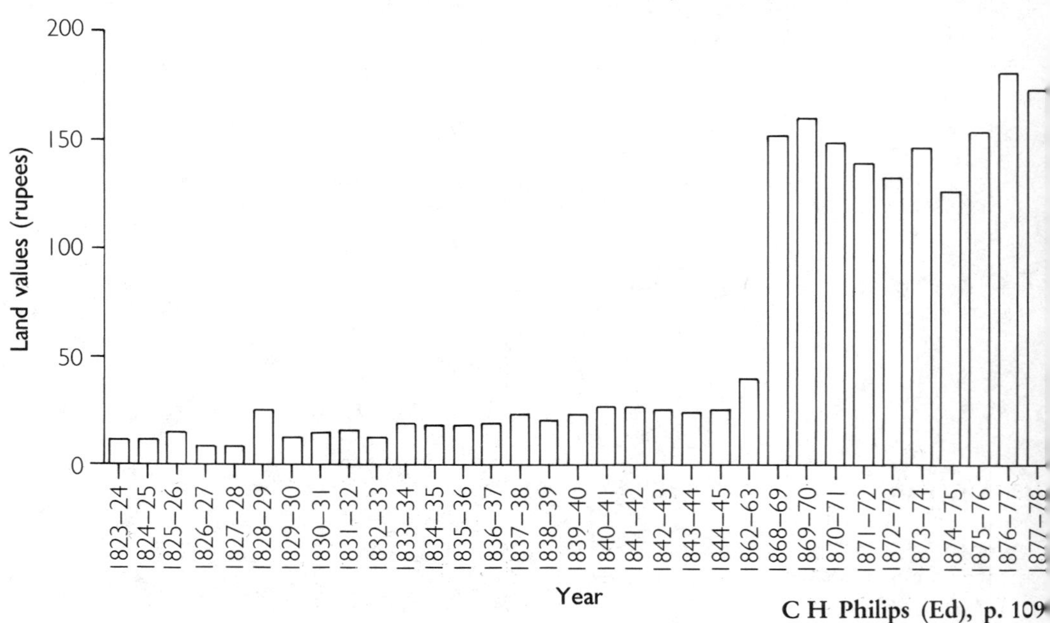

C H Philips (Ed), p. 109

Question • How would you explain this dramatic rise in land prices?

Source 4.9 Area under rice cultivation in lower Burma, from 1830 to 1931–35

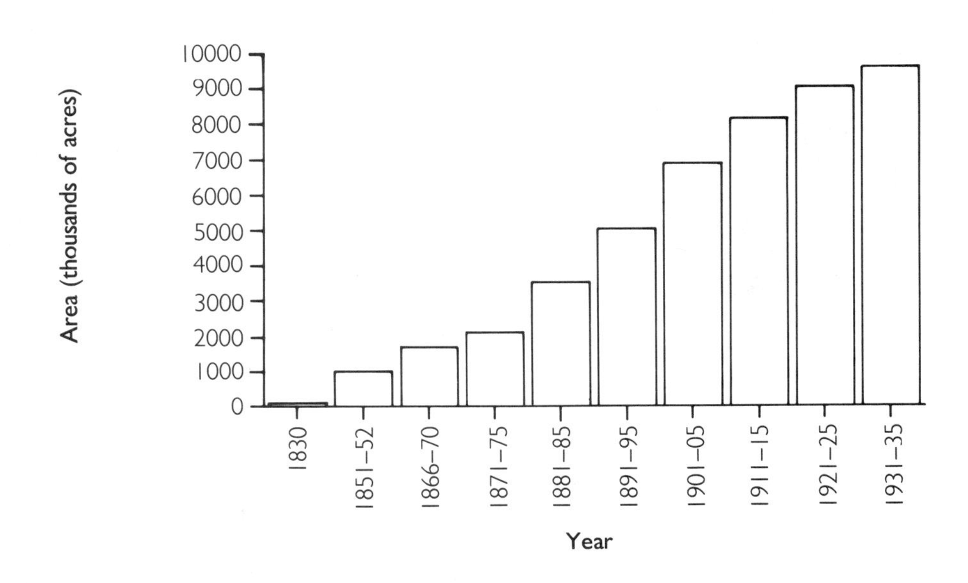

R von Albertini, *European Colonial Rule*, p. 120.

Question • The data portrayed in the bar graph suggest that for Burma the 19th century was an era of ever-increasing prosperity. Which classes in Burmese society would have profited most from the agricultural boom? Farm labourers? Share-croppers? Small landowners? Larger landowners? Grain merchants? Exporters? The colonial government?

Source 4.10 The contribution of the state to education in British India, 1901

There are other questions which I ask myself, and to which I cannot give the answer that I would like. I have remarked that we have been at work for 70 years. Even if we have done much, have we made the anticipated progress, and are we going ahead now? We are educating $4\frac{1}{2}$ millions out of the total population of British India. Is this a satisfactory or an adequate proportion? We spent upon education in the last year from public funds a sum of £1,140,000, as compared with £1,360,000 from fees and endowments. Is the State's contribution sufficient? Ought it to be increased? Is there an educational policy of the Government of India at all? If so, is it observed, and what is the machinery by which it is carried out?

C H Philips (Ed), p. 743.

Question • On these figures, would you regard the educational performance of the Government of India as 'satisfactory or adequate'?

Negative Aspects of Colonialism

The Summer Palace, Peking.

cultivation system Indonesian system in which peasants had to devote one-fifth of their land to produce export crops or provide the equivalent in labour in order to raise revenue and foster commercial agriculture

The colonial regimes definitely could have provided more in the way of social services for their subjects—who were also taxpayers. However, the main criticism that is now levelled at colonialism is not that it did too little, but that it did too much. Instead of making things better, it made them worse for the vast majority of Asian peoples. For instance, it is alleged that colonialism deprived Asians of their autonomy, robbed them of their treasures, taxed them excessively, and destroyed flourishing handicraft industries by flooding the market with cheap, factory-made products.

What evidence exists for these grave charges? At first sight there would appear to be plenty, and we give a sample below. In source 4.11, English historian Christopher Hibbert describes, with ironical understatement, the frenzied looting which occurred after the Anglo-French army captured the Chinese imperial Summer Palace outside Peking in 1860, during the second 'Opium War'. In source 4.12, the evils of the Dutch **cultivation system** in Java are catalogued in a report from L Vitalis, one of the officials whose duty it was to oversee the working of the System. At the time the report was suppressed. Source 4.14 is a statistical overview of the devastating impact of the Industrial Revolution on Asia's manufacturing sector. Again, this evidence must be interpreted cautiously. The fact that a few hundred soldiers behaved badly in Peking does not, by itself, prove that all colonials were crooks and vandals. Eventually the Dutch were moved by reports such as Vitalis' to discontinue the Cultivation System.

In source 4.13, Sir Henry Rawlinson offers what could be classed as a fairly reasonable justification for the 20 million pounds a year which India paid for the 'privilege' of British rule.

Source 4.11 The sacking of the Summer Palace of the Ch'ing Emperors, 1860

On 5 October, the siege-train having at last arrived from Tientsin, the armies began to move cautiously forward again, the French on the left, the British on the right. In the difficult country, cut up by brickfields and kilns, tombs and walled gardens, traversed by narrow, sunken roads, the two armies lost contact with each other, and the British, held up for some hours by Tartar cavalry, failed to reach the rendezvous at the Summer Palace of Yüan-ming Yüan suggested by Hope Grant for the night of 6 October. When the British vanguard did arrive there on the morning of the 7th they found that the French army, 'officers and men, seemed to have been seized with a temporary insanity; in body and soul they were absorbed in one pursuit, which was plunder, plunder'. . . .

When the British arrived they found the ground covered with lanterns and masks, broken boxes and wrapping paper, smashed porcelain, yellow cushions embroidered with the five-clawed dragon, Louis XV clocks smashed to pieces for the sake of their jewels, silks and clothing of all kinds. 'The men [were still running] hither and thither in search of further plunder, most of them . . . being decked out in the most ridiculous-looking costumes they could find. . . . Some had dressed themselves in the richly-embroidered gowns of women, and almost all had substituted the turned-up Mandarin hat for their ordinary forage cap. . . . One of the regiments was supposed to be parading; but although their fall in was sounded over and over again, I do not believe there was an average of ten men a company present.' . . .

When Lord Elgin arrived, de Montauban came up 'full of protestations. He had prevented *looting* in order that all the plunder might be divided between the armies, etc. etc.' He proposed that the spoils should be equally divided between the two armies, saw to it that Elgin himself received one of the Emperor's green jade batons, and that when a hoard of 800,000 francs worth of gold and silver ingots was discovered, the English got half. But, as Elgin could see, despite de Montauban's protestations, there was no possibility of there being any equal division of the spoils or any stop to the looting which was still proceeding apace. . .

According to Count d'Hérisson this was a mistake more often made by Frenchmen than the British. The British, indeed, he said, were far more methodical in the looting than his own countrymen, going about it in a highly organized manner. They 'moved in by squads, as if on fatigue duty, carrying bags, and were commanded by non-commissioned officers who, incredible as it sounds (yet strictly true), had brought jewellers' touchstones with them'.

One of their chaplains would have joined in, too, and taken one of the temple idols if 'that lazy syce' of his had been there to carry it off for him; but as it was it proved too heavy and he left it on the temple's marble floor where his companion had thrown it down to see if he could smash it. Another chaplain cut short the Sunday service to ride over to the palace from which he returned with a mule cart piled high with loot. The next Sunday he preached an 'admirable sermon against covetousness'.

C Hibbert, pp. 271, 273–5.

Question • In the extract Hibbert might be said to be using an ironical tone. How does this literary device add strength to his account?

Source 4.12 The 'Culture System' in the Priangan District, Java, 1835

The forced cultivation of indigo was introduced in the Priangan in 1830. And in 1835 I was charged with the inspection of the factories in this residency in order to determine whether the sorry state of affairs pictured by the Resident Holmberg de Beckfelt was not exaggerated . . . The following extract from my report of 29 January 1835 is indicative of the situation of these poor peasants:

> Truly their situation is lamentable and really miserable. What else can one expect? On the roads as well as in the plantations one does not meet people but only walking skeletons, which drag themselves with great difficulty from one place to another, often dying in the process. The Regent of Sukapura told me that some of the labourers who work in the plantations are in such a state of exhaustion that they die almost immediately after they have eaten from the food which is given to them as an advance payment for the produce to be delivered later. If I had not witnessed this myself I would have been hesitant in reporting this . . . These victims can even be found on the roads leading from Tasikmalaja, Garoet, Ardjawinangon, and Galo. One even passes them unnoticed! What then must be the fate of those who collapse on the desolate roads and paths? The Regent, when asked why he did not have the bodies buried, replied: 'Every night these bodies are dragged away by the tigers.' . . .

C L M Penders (Ed), p. 24.

Question • Why would a historian of the Culture System pay close attention to Vitalis' report?

Source 4.13 A rebuttal of the 'drain theory' by Sir Henry Rawlinson

The so-called 'drain,' or more specifically the Home Charges, consisted of money remitted to England to pay for the salaries and pensions of the civil and military officials employed in establishing law and order, and protecting the country from invasion, for the purchase of stores unobtainable in the country, and finally, for paying the interest on loans raised for the construction of irrigation works, roads and railways. The sum paid was small in proportion to the benefits received; it is idle to pretend that, at the time when India was taken over by the Crown, she could provide men of the type required to build up an orderly system of government on the ruins of the Mogul Empire.

The loans required by Dalhousie and his successors could not have been raised locally, and were obtained on particularly favourable terms. The average rate of interest paid by India before the [1914–18] war was 3½ per cent, whereas Japan could not obtain the money that she required for similar purposes at less than 5½ per cent. Indian loans were trustee stock, backed by the British Government. Moreover, it was a productive debt, as both irrigation canals and railways returned profits which left over a substantial margin when the charges had been met. To this must be added the British enterprise and capital invested in tea, coffee and rubber plantations, which gave employment to many thousands of workers. Indian exports rose from £8 millions in 1834 to £250 millions in 1928.

H G Rawlinson, p. 234.

Question • Do you find Rawlinson's argument that the Indians got value for money from British colonialism convincing?

Source 4.14 Distribution of world manufacturing output with special reference to India, 1750–1938

Year	*India*	*Other Third World countries*	*Developed countries*
1750	25	48	27
1800	20	48	32
1830	18	43	40
1860	9	28	63
1880	3	18	79
1900	2	9	89
1913	1	7	93
1928	2	5	93
1938	2	5	93

Adapted from P Bairoch, 'International Industrial Levels', pp. 296–304, and C Simmons, 'De-industrialisation', pp. 593–622.

Question • This table paints a very bleak picture of the Indian economy. But do the figures actually prove that it was going downhill?

Colonialism and Christianity

Earlier, it was suggested that prosyletisation was one of the main initial reasons for Europe's expansion into the East. However, the Spanish and Portuguese zeal for conversion was not generally matched by their Protestant successors—the Dutch, Danes, English and Americans. Alexander Duff (source 1.15) exemplifies the Protestant churches' eagerness to take the word of God into the East.

From the late 18th century these churches put much time and money into organising missionary societies such as the Church Mission Society (Anglican) and the American Board of Foreign Missions. But, on the whole, these societies received little support from colonial governments which resented their interference and feared its possible political consequences.

elling Christian
terature, Chung King,
China.

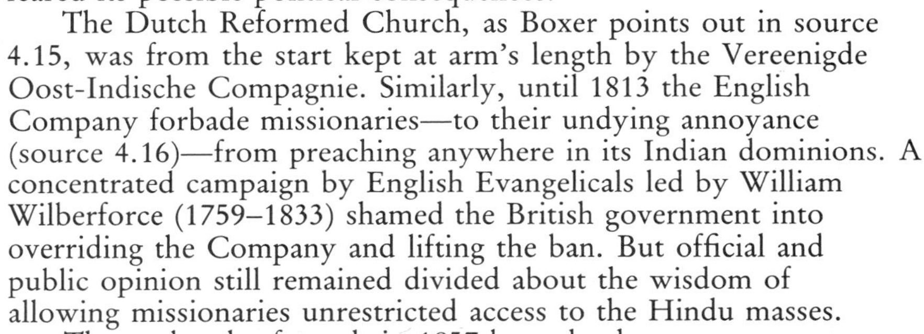

The Dutch Reformed Church, as Boxer points out in source 4.15, was from the start kept at arm's length by the Vereenigde Oost-Indische Compagnie. Similarly, until 1813 the English Company forbade missionaries—to their undying annoyance (source 4.16)—from preaching anywhere in its Indian dominions. A concentrated campaign by English Evangelicals led by William Wilberforce (1759–1833) shamed the British government into overriding the Company and lifting the ban. But official and public opinion still remained divided about the wisdom of allowing missionaries unrestricted access to the Hindu masses.

The outbreak of revolt in 1857 brought the controversy to boiling point. Source 4.17 cites first an anti-missionary outburst from *The Illustrated London News*, and then a defence of proselytisation from one of the most Christian of 19th century administrators, the Chief Commissioner of the Punjab, (Sir) John Lawrence (1811–79).

Source 4.15 Charles Boxer on the status of the Dutch Reformed Church in Asia

The extent to which the Dutch Reformed Church in Asia was subordinated to the civil power was strikingly emphasized in 1653. In October of that year the Governor-General and Council ordered a thanksgiving prayer- and fast-day to commemorate Dutch victories against the 'rebels' in the Moluccas and to pray for the further success of Dutch arms. The Batavia church council ventured to criticize this resolution, on the grounds that the war in Amboina was not a just one, having been provoked through the tyrannical misbehaviour of the Dutch themselves—as indeed it had. This remonstrance drew a stinging rebuke from the Governor-General, who accused the church councillors of being highly unpatriotic and of 'giving a bad impression of the Company's righteous trade'. The directors at Amsterdam reacted even more violently, and ordered that any *predikanten* [ministers of Dutch Church] who ventured to make similar criticism in the future should be dismissed immediately from their posts and embarked for the Netherlands in the first available ship. Before this dispatch reached Batavia, the local church council had come crouchingly to heel at the summons of the Governor-General, and the two *predikanten* who had instigated the critical motion escaped deportation by an abject apology and retraction. So far as I am aware, the Company experienced no further criticism from the ministers of the Dutch Reformed Church concerning its wars, whether just or unjust, for the remainder of its existence.

C R Boxer, p. 154.

Question • Can you think of a more recent case in which the Dutch Reformed Church has been accused of playing politics?

Source 4.16 An appeal by the American Board of Missions to the Governor of Bombay, *c.* 1812

We entreat you by the spiritual miseries of the heathen, who are daily perishing before your eyes, and under your Excellency's government, not to prevent us from preaching Christ to them. We entreat you by the blood of Jesus which he shed to redeem them,—as ministers of Him, who has all power in heaven and earth, and who with his farewell and ascending voice commanded his ministers to go and teach all nations, we entreat you not to prohibit us from teaching these heathens. By all the principles of our holy religion, by which you hope to be saved, we entreat you not to hinder us from preaching the same religion to these perishing idolaters. By all the solemnities of the judgment day, when your Excellency must meet your heathen subjects before God's tribunal, we entreat you not to hinder us from preaching to them that gospel, which is able to prepare them, as well as you, for that awful day.

P D Curtin (Ed), *Imperialism*, p. 220.

Question • Although framed as an appeal, the Board's letter contained a barely disguised threat. What was it?

Source 4.17 Two views on Christianity and the verdict of the Indian mutiny, 1857–58

a We owe the people of India much. We owe them peace, we owe them security, we owe them good government; and if we pay them these debts many blessings will follow. . . . Let us not make the mistake of thinking that we owe them Christianity, and of attempting to force it upon them before they are ripe to receive it. Christianity was never yet successfully inculcated [impressed] by the sword, and never will be.

***The Illustrated London News*, 4 July 1857.**

b I believe that, provided neither force nor fraud were used, Christ would assuredly approve of the introduction of the Bible. We believe that the Bible is true, that it is the only means of salvation. Surely we should lend our influence in making it known to our subjects. A Turk who acted up to his own convictions would only act consistently in inculcating the study of the Koran. But whether he acted rightly or wrongly in doing so, is for a higher Judge to determine. As a matter of policy I advocate the introduction of the Bible, quite as much as a matter of duty. I believe that, provided we do it wisely and judiciously, the people will gradually read that book. The whole question seems to me to resolve itself into what is the just interpretation of the term 'toleration.' I consider that it means forbearance;

that is to say, that we are to bear with and not to persecute mankind for their religious opinions. But this cannot mean that we should not strive by gentle means to lead those in the right way whom we see to be going wrong.

R Bosworth Smith, p. 313.

Questions

- Why was the *News* so hostile to the notion of imposing Christianity on India?
- How does Lawrence attempt to reconcile his obligations as an administrator and a practising Christian?

Unplanned Consequences of Colonial Policies

Under colonialism, change was probably more rapid than at any previous time in Asian history. However, not all the environmental and social changes that took place were the direct consequence of colonial policies. Some were brought about by the inflexible laws of market economics, and others were initiated by Asians as part of a continuing process of cultural reappraisal. But some were the result of badly thought out colonial policies gone wrong, which this section considers, with examples from the fields of justice, public works and local administration.

Justice was something in which all colonial regimes legitimately took pride. Unlike most of their predecessors, the European rulers dispensed justice incorruptly and with regard to firm rules of procedure. Theirs was undoubtedly a rule of law. This impartial but wooden system was alien in many ways to Asian traditions, and often broke down when confronted with the deeply rooted prejudices of ancient custom. Source 4.18, from the memoirs of a late 19th century Bengal civilian Sir A H L Fraser, shows the potential for conflict between Western justice and Eastern tradition.

autonomous having personal freedom of will and action, independent

The Westerners were also justly proud of their centralised, bureaucratic system of administration which united thousands of hitherto locally **autonomous** villages. Source 4.20 is from an essay by Dutch official B J Haga. It makes clear that administrative centralisation, which altered the position of the village headman, did not necessarily lead to a more sympathetic and responsive mode of government.

'he Quilon express to Quilon halts at a station.

And if Westernising the administration caused some problems so, too, did the development of the countryside. Roads, railways and irrigation canals were probably the most substantial achievements of Western colonial rule in Asia. But their development had unintended consequences. For instance, in source 4.19 American historian Ira Klein explains how the building of canal, road and railway embankments in Bengal created stagnant pools which (unknown at the time) were perfect breeding places for the anopheles mosquito, the carrier of the malaria bacillus.

Source 4.18 Law versus custom in British India

As a young officer, I was once called on to inquire into a case of murder and to prosecute it before the Court of Session. It was as clear a case as ever had been. The murder was cruel, and the eye-witnesses were beyond suspicion. There were two assessors [jurors], and both of them returned a verdict of 'not guilty.' The Judge differing from the assessors sentenced the Brahman accused to death, and he paid the penalty. Some time afterwards, one of the assessors came to visit me. He was a fairly influential landowner, and himself a Brahman, well educated in the vernacular, but without knowledge of English. I asked him how he could find a verdict so contrary to the evidence, and he frankly said to me in the most friendly way, 'I could not possibly find a verdict which would lead to the death of a Brahman. You know that it is grievous sin for any Hindu to cause the death of a Brahman; and it does not matter whether you do it with your own hand or indirectly by the hand of another.' 'But,' I said, 'it is a serious thing for you to betray the trust which is reposed in you by the Government on behalf of the public; and you cannot help regarding this as most blameworthy failure of duty.'

He replied with some emotion, 'It is you really who are to blame. You are not ignorant of our views in this matter. Why, then, should you put us in a position where we might be called upon, as I was on that occasion, to choose between the sin of saying what I believed to be untrue, and the infinitely awful sin of causing the death of a Brahman?' The strong feeling with which my old friend spoke to me on the subject made a great impression on my mind, and I have often thought that we do not know, or at all events do not fully consider, what grievous injury we inflict on the people of India by forcing on them customs and duties which are altogether inconsistent with their traditions and beliefs.

A H L Fraser, pp. 340–2.

Question • Was justice done in this case?

Source 4.19 Roads, railways and malaria in Bengal, 1840–80

The abruptly rising level of admissions to thana dispensaries for intermittent fever told grimly the story of the advances of malaria, particularly in Western and Central Bengal. In the 1840's fever admissions had been no higher, and in some instances lower, in Burdwan, Hooghly, and elsewhere in these regions than for the relatively healthy eastern district of Dacca. In Hooghly, the rate was under five per cent; in Midnapore about five or six per cent; in Jessore, one of the earliest districts to become malarial, fever admissions remained under ten per cent at mid-century. As late as 1873, the spleen index at Burdwan varied from a low of .4% at Bood Bood to a high of 7.8% at Shodpur. But by 1900, the spleen rate through the district ranged from 35% to ninety per cent. In Midnapore, fever cases had doubled by the 1860's and passed fifty percent by the late 1870's. In Jessore they passed forty per cent by 1880. In Birbhum they rose from eight per cent in 1865 to sixty per cent in 1880 . . .

Why did malaria proliferate with such terrible effect through large parts of Bengal in the last half of the nineteenth century? As early as the 1840's British health officials and engineers and some Indian commentators had a notion that malaria spread with works of economic development. The first systematic inquiry appears to have been conducted by a canal committee appointed in the mid-1840's to explore the causes of the spread of unhealthy conditions around the Delhi and Jumna canal works. T E Dempster, a surgeon with the Horse Artillery, and a member of the canal committee, cited the 'astonishing difference' between the number of fever cases belonging to the 'irrigated and unirrigated parts' of the region of the West Jumna canal. The canal committee found that the spread of malaria was 'intimately connected with this canal construction in a remarkable degree.' They discovered two remarkable local conditions common to all tracts irrigated from the existing canals where spleen disease prevails to a great extent, viz. obstructed surface drainage and a stiff retentive soil. Where soil conditions and drainage were favorable malaria had not taken hold. . .

Similarly, road building was linked intimately with the proliferation of malaria. This fact began to be recognized by a few in the 1860's. In the early nineteenth century, Bengal, for example, remained without railroads and was almost devoid of roads. There were in Jessore 'only twenty miles of roads,' and 'not one hundred carts' in 1800. In 1837 there was 'not a single road in Hooghly which a European vehicle could traverse,' and as late as 1870 there were 'no roads and not a single cart in Malda.' One of the first known malaria epidemics began in Jessore in 1836 and was 'associated with the construction of a road.' By 1863 the deadly outburst of malaria in large parts of Bengal led to the appointment of a government committee of inquiry. Its Indian member, Raja Digambar Mitter, and others, noted the tie between malaria and road construction. Mitter was the 'chief exponent' of the embankment theory, that railway and road embankments stimulated malaria by causing water logging. . .

When modern transport works were completed, increased mobility further transmitted disease around the country. River boat crews frequently were malarial, and third-class railway waiting rooms were notorious centers of infection.

I Klein, pp. 138–40, 142–3.

Question • Is this writer suggesting that the roads and railways which helped to spread disease should not have been built?

Source 4.20 The changing role of the village headman in Dutch Borneo

In the first place the headman charged with the execution of the task of the Government felt the authority of the Government behind him; as a result he had less and less the feeling that he should discuss matters with the members of the native council and that he should act in accordance with the opinions of the population. More or less, he managed, with the authority of the Government behind him, to acquire a more autocratic position; this was also due to the fact that the native safety valves and means of defence, which the native community previously possessed in respect of their rulers, had disappeared or become less effectual owing to the presence of the Governmental authority.

The judgment of the manner in which a chief fulfilled his duties was transferred from the people to the Government, which used a different standard to that of the people. The Government was naturally inclined to relieve a headman of his duties who, whilst looking after the interests of his community, did not carry out the task of the Government to its satisfaction; on the other hand it was not quick to dismiss a headman who performed his duties towards the Government in a satisfactory manner but who, in the eyes of the people, fell short of his native obligations.

B J Haga, p. 185.

Question • Was the position of the village headman in Borneo strengthened or weakened by the advent of Dutch rule?

5 *Asians as Collaborators*

If, however, security was wanting against popular tumult or revolution I should say that the Permanent Settlement . . . has thi great advantage at least of having created a vast body of rich landed proprietors deeply interested in the continuation of British Dominion . . .

Lord William Bentinck, 182

IT IS DIFFICULT TO imagine that Asians could enjoy being ruled over by foreigners. Yet one of the more intriguing aspects of the colonial story in Asia is that Western rule was, at least to begin with, favourably received by the majority of its 'consumers'. The present chapter seeks to resolve this paradox by showing how colonial rule met the economic, social, and political needs of a variety of élite and upwardly mobile groups in both city and country. This and the chapter that follows together highlight the experience of Asians living in a world dominated by European power and culture.

KEY DATES

1793 The 'Permanent Settlement' of the Bengal land revenue introduced by Earl Cornwallis (1738–1805) gives legal recognition to the property rights of the *zamindars* (landlords).
1817 Outbreak of war between the English East India Company and the Maratha Confederacy led by the Peshwa of Poona.
1825 Sultans of Surakarta and Jogjakarta refuse to join Pangeran (Prince) Dipanagara's revolt against the Dutch in eastern Java.
1834 'Agency' firm of Carr, Tagore and Company founded in Calcutta.
1851 Beginning of the 17 year reign of King Mongkut Rama of Siam (Thailand).
1855 Treaty of friendship signed between Siam and Britain.
1865 Bengali Brahmin Satyendranath Tagore (b. 1842) becomes the first native-born member of the Indian Civil Service.
1874 Sir Andrew Clarke (1824–1902) negotiates the Pangkor Engagement with the chiefs of Perak (Malaysia).
1888 United Indian Patriotic Association founded by Muslim loyalist Sir Saiyyid Ahmad Khan (1817–98).
1893 Anglo-French Convention underwrites the independence of the remainder of Siam.
1898 Filipino guerilla leader Emilio Aguinaldo (1869–1964) hails the United States for 'liberating' his country from Spanish rule.
1914–18 Twelve Indian soldiers win the Victoria Cross during World War I.

ISSUES AND PROBLEMS

The relative ease with which the Europeans maintained their control of large native populations has always constituted one of the most intriguing aspects of the colonial story in Asia. 'The essential issue in all this', writes D A Low, 'can, I think, be illustrated very simply with just one question. How was it that 760 British members of the ruling Indian Civil Service could, as late as 1939, in the face of the massive force of the Indian national movement led by Gandhi, hold down 378 million Indians?'[1] Nowadays, most scholars would say that the key factor was the readiness of many colonial subjects to *collaborate* with—that is, actively assist—their foreign masters.

Eurasians persons of mixed European and Asian blood

In what ways did Asians 'collaborate'? Most obviously, they collaborated by taking employment with colonial governments or European firms. The British Raj might have been, in spirit, a foreign regime, but at the lower levels especially it was manned primarily by Indians and **Eurasians** (source 5.4). In 1885, for example, nearly two-thirds of its 20 000 administrative and clerical positions were held by natives. Similarly, in the later 19th century, two of every three places in the non-commissioned ranks of the Indian Army were held by natives. Indeed, government service often ran in the family. Do-Huu-Phuong, a prominent Vietnamese collaborator under the French, sent one of his sons to the army, another to the air force, and two more to serve in the government of Cochin China. As far as private employment is concerned, it is fair to say that had it not been for the invaluable assistance rendered by Asian middlemen—the *dragomen* of Turkey, the *dubashes* of India, and the *compradores* of China—European commercial interests could not have penetrated Asian markets to the extent that they did.

Second, natives collaborated by identifying publicly with the colonial regime. During the Indian Mutiny, for example, large parts of north India passed temporarily out of British hands. Yet even in the areas that were not affected, many prominent Indians stood by the British, helping them to escape the rebels and protecting their property from being looted. In Calcutta, a public meeting attended by most of the city's gentry passed a unanimous vote of support for the success of British arms. The princes, too, generally remained loyal, earning a commendation from the Viceroy, Earl Canning, for being 'breakwaters in the storm'. In addition, natives routinely registered their allegiance by accepting government favours. In Indonesia, members of the aristocracy were incorporated into the Dutch administration as 'regents'. In Burma and Ceylon, ambitious natives put themselves up for election to legislative councils. In Vietnam, educated mandarins competed for the rare 'privilege' of French citizenship. And in India, wealthy landlords and businessmen queued up at Government House to be invested with imperial honours. Gandhi launched his first campaign of 'non-co-operation' with the Raj in 1920. Significantly, one of his first acts was to urge his countrymen to renounce their imperial titles. By January 1921, only 24 of 5186 titleholders had done so. Once begun, the habit of collaboration was not easily broken.

Last but not least, Asians voted for colonialism with their feet, migrating in their millions during the 19th and 20th centuries to the new coastal cities and other areas under European administration. Calcutta was a fishing village when it was purchased by the English East India Company in 1690. By 1900 it had grown into a teeming metropolis of 1.2 million, making it the second largest city of Asia (Tokyo was the largest). Similar explosive growth was experienced by Bombay, Rangoon, Batavia and Saigon. However, the attraction of colonial rule was not confined to the ports; rural areas also prospered. At the time of British annexation in 1824, the southern Burmese provinces of Arakan and Tenasserim had populations of 100 000 and 70 000 respectively. By 1855 their combined population was 560 000, that is, the population had more than tripled in 30 years. Clearly, many Burmese felt British rule was preferable to that of the Ava Dynasty which was still in control in the north of the country.

As a rule, history does not look kindly on collaborators. In general, modern writers have portrayed the Asian élites who supported colonial rule as traitors who 'sold out' their countries. Nevertheless, in order to understand how colonialism worked in Asia, we need to know why these Asians acted as they did. What considerations drove literally millions of people over the last two centuries to co-operate with foreigners?

THE DEBATE

Broadly speaking, there are two schools of thought on this issue. The first, represented by people like Low and Ronald Robinson, believes that Asians collaborated out of self-interest. Low writes that

> there were always some indigenous peoples in a colonial society who had a vested interest in its maintenance. In the first place these were composed of those who were enjoying the profits of steady pursuits—perhaps teachers, perhaps traders, perhaps even peasants. But, as has often been pointed out, there were in addition always some indigenous peoples in an imperial situation who were directly dependent for their positions of prestige, and for their jobs, upon the patronage of the imperial power; and in the nature of things these constituted yet another support for imperial authority.[2]

materialist tendency to prefer material possessions and physical comfort to spiritual values

Other scholars, however, do not find this somewhat **materialist** view of human motivation very compelling. For them, the bond between rulers and ruled in Asia was not money, status, or self-preservation but a shared system of values and beliefs. According to this viewpoint, Asians living in an urban context collaborated with colonialism because, despite its faults, it stood for reason and progress, and represented a vehicle of modernisation. (In a modern society some or all of the following features are present: (i) a widespread belief in progress, reason, logic and the scientific method; (ii) separation of church and state; (iii) opportunities for social mobility; (iv) a high degree of urbanisation and economic specialisation; (v) individual freedom; (vi) centralised government; and (vii) equality for all before the law.)

The European Peace

mercenary hired; working only for money or other reward

First and foremost among the material benefits which Western colonial rule brought to Asia was law and order. This enabled their subjects to go about their business unhindered by the threat of **mercenary** rulers, bandits, and *thugs*.[3]

It is true that peace was not enforced overnight, and many people would say that taxes became heavier rather than lighter under the more rigorous European system. Nevertheless, the colonial period compares favourably with the violent, anarchic conditions which prevailed in many parts of Asia before the coming of the Europeans.

According to long-time British administrator Sir Frank Swettenham, everyday life in pre-modern Malaya was characterised by slavery, random taxation and the abduction of women (source 5.1). Sir Saiyyid Ahmad Khan was an eminent Muslim scholar-official; in his contemporary history of the Indian Mutiny (source 5.2), he testifies to the dramatic improvement which British rule had wrought in the quality of life in north India—the *Pax Britannica*.

These documents indicate clearly that many Asians saw the Europeans as liberators from oppression. This might be difficult to believe today, given the current disrepute of colonialism, but a case in point is source 5.3, the proclamation by rebel leader Emilio Aguinaldo (1869–1964) welcoming American intervention in the Philippines.

Source 5.1 Political conditions in Malaya before the advent of British rule

In each State the ruler, whether he were sultan, raja, or chief of lower rank, was supreme and absolute. His word was law, and oppression and cruelty were the result. Under the ruler were a number of chiefs, usually hereditary, who took their cue from their master and often out-Heroded Herod in the gratification of their vengeance or the pursuit of their peculiar amusements. The people counted for nothing, except as the means of supplying their chiefs with the material for indulging their vicious tendencies. They occupied land, but they did not own it; they worked by command and without payment; they were liable to be deprived of anything they possessed that was worth the taking, or to be taxed to meet the necessities of the ruler or the local chieftain; their wives and daughters were often requisitioned by members of the ruling class, and when they ceased to any longer attract their abductors, these women, often accompanied by other members of their families, went to swell the ranks of the wretched 'debt-slaves,' a position from which they probably never escaped, but, while they filled it, were required to perform all menial duties and were passed from hand to hand in exchange for the amount of the so-called debt, exactly like any other marketable commodity. The murder of a *raiyat* [peasant] was a matter of easy settlement, if it ever caused enquiry, and for the man who felt himself oppressed beyond endurance, there was left that supreme cry of the hopeless injured, which seems, with the Malay, to take the place of suicide—I mean the blind desire to kill and be killed, which is known as *mĕngámok*. That was how the Malays were treated in their own country, and you will readily understand that the Chinaman was regarded as fair game, even by the Malay *raiyat*, who if he met a Chinaman on a lonely road (and nothing but jungle tracks existed) would stab him for a few dollars, and rest assured that no one would ever trouble to ask how it happened.

I have not exhausted the catalogue of horrors, I have only generally indicated some of them, they still exist upon our borders in the States of Trengganu and Kelantan, where as yet Malay methods of government prevail; but I have told you enough, and it is surely something to be able to say that, in every State where there is a British Resident, slavery of all kinds has been absolutely abolished; forced labour is only a memory; Courts of Law, presided over by trustworthy magistrates, mete out what we understand as justice to all classes and nationalities without respect of persons, and the lives and property of people in the protected Malay States are now as safe as in any part of Her Majesty's dominions.

P H Kratoska (Ed), pp. 177–8.

Question • Why would one want to question Swettenham's picture of traditional Malay society?

Source 5.2 An Indian Muslim defends British rule in Bijnor, *c.* 1860

A man should think about those events which happen in the world and strive to instruct himself from a study of their consequences. The turmoil of violence which happened was only a punishment for the ungratefulness of the Hindustanis. There are many men today who have experienced only English rule in their lives. They not only were born under English rule, but they came to maturity under it. In short, the sights which they saw were exclusively the sights of English rule and not of any other. In Hindustan, people are not at all accustomed to learn about former times from the facts of history, nor from reading books. It is for this reason that you people were not acquainted with the injustice and oppression that used to take place in the days of past rulers. Whether rich or poor, a person in those times could never be at ease. If you had been acquainted with the injustice and excesses of those past days you would have appreciated the value of English rule and given thanks to God. But you were never grateful to God and remained always discontented.

In the interval between the time when the Nawabs were in the ascendent [and the government regained control], the Hindus became strong enough to dominate for a few days, so that the Hindu landlords were able to rule the District. You could then taste the Government of Hindus and see what Muslims experienced at Hindu hands: how many houses looted, how many villages razed, and even your own womenfolk ravaged. Speak truthfully then. The English ruled fifty four years in this District. Did any person, Hindu or Muslim experience any trouble or annoyance then? Can you not recollect what were the disasters brought down on your head by this Hindustani Government in those times of the rebellion? Go to the history books on the rule of the former great emperors to measure the extent of the cruelties and disasters borne by the common folk of those days of organized governments.

Not even one hundred-thousandth part of the ease was present then which fell to your lot in English times. Look how Hindus and Muslims are living with all ease and in peace under English rule. The strong cannot tyrannize the weak now. Each worships God and his Creator according to the requirements of his religion. There is an atmosphere of live-and-let-live. The Hindu builds temples in which to worship; the Muslim builds mosques where prayers are read and the call to prayer is uttered. There is no one to stop and not one to forbid. The merchant pursues his trading affairs, entrusting goods worth thousands to an infirm and aged agent, who is sent thousands of miles to earn a profit; and there is no fear of dacoit [member of Indian armed robber band] or thug. And roads—how perfectly secure they are; women, adorned with jewelry worth thousands of rupees, may ride at night in horse-drawn carriages from stage to stage, all quite free of anxiety. The owner-cultivator is busy in the fields; no one takes an iota more than the appointed rent for these fields.

S Ahmad Khan, *Bijnor Rebellion*, pp. 107–8.

Question • Why might Sir Saiyyid Ahmad Khan be considered a biased witness?

Source 5.3 Emilio Aguinaldo's proclamation from Cavite, 1898

The great North American nation, a lover of true liberty, and therefore desirous of liberating our country from the tyranny and despotism to which it has been subjected by its rulers, has decided to give us disinterested protection, considering us sufficiently able and civilized to govern ourselves.

In order to retain this high opinion of the never to be too highly praised and great nation of North America, we should abominate such acts as pillage and robbery of every description, and acts of violence against persons and property.

To avoid international complications during the campaign, I decree:

1. Lives and property of all foreigners are to be respected, including Chinese and those Spaniards who neither directly nor indirectly have taken up arms against us.

2. The lives and property of our enemies who lay down their arms are to be equally respected.

3. In the same way, all hospitals and all ambulances, together with the persons and effects therein, as well as their staffs, are to be respected, unless they show themselves hostile.

4. Those who disobey what is set forth in the three former articles shall be tried by summary courtmartial and shot, if by such disobedience there has been caused assassination, fires, robbery, or violence.

House Documents, **55th Congress, No. 3, p. 103.**

Question • Why did Aguinaldo assume (wrongly) that, having defeated Spain, the United States would allow the Filipinos to rule themselves?

Native servants of the British Empire: Malay Armed Police (top left), Hongkong Chinese Police (bottom left), and Hongkong Sikh Police (right).

Prestige of Government Service

As well as appreciating the comparative even-handedness of European rule, Asians welcomed the opportunities that it afforded for regular employment in the army and bureaucracy. By 1885, Indians comprised half of the staff of the British Raj (source 5.4). Not unexpectedly, Asians were excluded from the top posts which carried real responsibility and high salaries until fairly late in the history of colonialism. Under new policies designed to prepare Asians for self-government, such posts did become available. They were keenly sought, despite being declared 'black' by the nationalists.

Why did so many collaborate? For money? For the high status that went with a government job? Doubtless these were important factors. It would be simplistic, however, to suggest that only greed caused Asians to collaborate with their European overlords. For instance, the rank-and-file foot soldier was rather poorly paid, and subject to the most formidable discipline, yet the vast majority of *sepoys* served their officers with faithfulness and gallantry. In his moving description of the Battle of Koregaum fought on New Year's Day 1818 near Poona in western India,

Philip Mason shows that this loyalty often went far beyond the call of duty (source 5.5).

Likewise, native domestic servants frequently developed a real affection for their masters. The encounter described in source 5.6 would be unremarkable had it occurred during the lifetime of the Indian Empire, but it did not. It occurred in November 1987, overheard by a travel-writer for the Melbourne *Age*.

Source 5.4 Racial distribution of senior government posts in British India, 1867 and 1887

Year	*Europeans*		*Eurasians*		*Indians*	
	Number	*%*	*Number*	*%*	*Number*	*%*
1867	4760	35	2633	20	6008	45
1887	6154	29	4164	19	11148	52

A Seal, p. 360.

Question

- Why do you think Eurasians (a small minority of the population) were so strongly represented in government?

Source 5.5 The battle of Koregaum, India, 1818

What made Indian soldiers give their lives for a flag they could hardly call their own? National pride did not play much part till late in their long history. It was only in the Second World War that it appeared and then only occasionally. When it did, it was a two-edged sword: pride in the regiment, in the division, yes, that was something on which everyone could agree, but pride in a nation that was not yet a nation produced very mixed feelings. Officers and men could not share it in the same spirit. Was it then for money that they fought? Certainly their pay, though far from princely, was relatively better than the British soldier's. But it explains nothing to say simply that they were 'mercenary'. Men may come to the colours for pay but it is not for pay that they earn the Victoria Cross. It is an unprofitable servant who does only his duty and there are times when a soldier is expected to do far more. These men often did.

Take, for example, the affair at Koregaum on New Year's Day, 1818. Captain Staunton, of the Bombay Army, received orders on New Year's Eve to bring the troops under his command to reinforce Colonel Burr, 41 miles away, whose two battalions were in danger of being cut off by the larger forces of the Peshwa, the Maratha chief. Staunton marched at eight o'clock that evening. He had some 500 sepoys, that is, privates, of the Bombay Army (the 2nd Battalion of the 1st Bombay Regiment, later the 2nd Battalion, 4th Bombay Grenadiers), 250 newly raised auxiliary horse (who eventually became the Poona Horse), and a party of the Madras Artillery. There were two six-pounder guns with 24 European gunners and some Indian drivers and gun lascars, the whole artillery contingent probably numbering about 100 men, the entire force being less than 900. They marched all night, covering 27 miles, and at about ten in the morning on New Year's Day were preparing to halt for rest during the heat of the day, when they came suddenly in view of the Peshwa's main army. This consisted of about 20,000 horse and 8,000 infantry. The village of Koregaum, with stone-built houses and enclosures, was close at hand,

and Staunton at once saw that his best move was to take up a defensive position in the village. But the enemy were equally quick and dispatched three columns of their best Arab troops, each column about a thousand strong, to prevent him.

Neither side was able to occupy the whole village before the other, and there developed a battle, house to house and hand to hand, in the course of which almost every building was taken and retaken. The horse could not be used as cavalry; they fought on foot as infantry. It was thus an infantry and artillery battle, centring on the two guns . . . About nine o'clock that night, twenty-five hours after their march had begun, Staunton's force cleared the village of the enemy and were able to get water.

When it was light, they found that the whole enemy army was withdrawing, doubtless because they had news of British forces moving up to join Colonel Burr. Staunton's force brought away colours, guns and wounded and marched for the nearest garrison town, where they arrived forty-eight hours after they had left their own station. During that time they had had no food and had been constantly marching or fighting. For nearly twelve hours of fighting they had been without water. The 2nd/1st had lost 50 men killed and 105 wounded out of fewer than 500. Of the 250 auxiliary horse, ninety were killed, wounded or missing. Of nine British officers, only Staunton, one subaltern and the doctor were unwounded; also four gunners out of twenty-four. But, before reaching the town where they might expect rest, they halted to dress their ranks and marched in with drums beating and colours flying.

Many such tales could be told, and the question is still unanswered. Their general wrote, after Koregaum, of the sepoys' 'most noble devotion and most romantic bravery under pressure of thirst and hunger almost beyond human endurance'. Why did they give such service? Was it for a naked oath—the oath of allegiance they swore when they passed out as recruits and were enlisted as soldiers? Was it the personal honour of a man who followed the profession of arms, as his ancestors had done and as his descendants would do after him, and who feared the shame of cowardice? Was it respect—sometimes even affection—for the personal qualities of his officers? Was it fear of an iron discipline? Was there simply a long-instilled habit of endurance? What were the links in the chain of confidence and why did they sometimes snap?

P Mason, pp. 15–17.

Questions

- Why did the *sepoys* who served at Koregaum have no 'national pride'?
- Where else could their loyalties have lain?

Source 5.6 A master and servant remember the old days of the Ootacamund Club, 1987

Then an elderly gentleman entered. The butler seated him two tables away. Immediately, the years of Indian independence vanished. The Raj lived. 'Sir John, we have missed you in the club', the Butler said. 'It's been some time, 22 years actually. Good to see Ooty hasn't gone to pot', the old gentleman replied. 'We try to maintain the real standards. And my condolences. The brigadier told us of your sad loss.' 'Thank you butler. Memsahib often mentioned you with appreciation when we chatted about the old days.' 'The usual sir?' 'Thank you, the usual.' Without a further word the butler brought a double scotch.

The *Age Extra*, 28 November 1987.

Question

- What is meant by the expression 'gone to pot'?

Supporting Role of Asian Élite

patronage support, encouragement given by benefactor

protected states Asian states protected by Europe but remaining internally self-governing

His Highness The Maharaja of Jammu and Kashmir.

amnesty general pardon, especially for political offence

Various classes of Asians also benefited from the protection and **patronage** of the colonial regimes. One was the aristocracy. The Europeans found it convenient to leave some of the native princes on their thrones and to control their states indirectly through the agency of residents attached to their courts. In the process, the princes lost some of their power. At the same time they gained what was perhaps—for men who valued pomp and display—ultimately more important, complete security of position. Their fragile thrones were henceforth guaranteed by imperial military power, their kingdoms becoming **protected states**. For example in the treaty signed with the chiefs of Perak State in 1874 (source 5.8), Raja Abdullah was recognised as Sultan and assured of full British protection; all he had to do in return was to abide by the stipulations of the treaty.[4]

Like the princes, the large rural landowners were regarded by the Europeans as 'natural leaders' of the people and therefore as useful allies. These landowners also profited by aligning themselves with the Europeans: they received physical protection from the colonial police and military forces, and the viability of their estates was maintained by the colonials' economic policies. In his role as Commissioner of Lucknow, Colonel Barrow describes the fate of the Talukdars of Awadh (Oudh) after the Indian rebellion of 1857 was suppressed (source 5.7). The Talukdars, a group of rural aristocrats, were persuaded to lay down their arms only after promises of an **amnesty** and the restoration of lands previously taken away from them. Later in the 19th century they gained imperial honours, appointments to the bench as magistrates, and nominations to the United Provinces Legislative Council.

Source 5.7 Colonel Barrow on the political motives behind the restoration of the Oudh Talukdars, 1866

5. After the fall of Lucknow the land-owners, with but few exceptions, still held back. Sir Robert Montgomery, under Lord Canning's confiscation order of March, held the proprietary rights of the whole province in his hands. My instructions simply were to use this powerful lever to induce powerful men to become the friends and supporters of Government and good order; and under the arguments then used, about two-thirds of the Talookdars tendered their allegiance before the army took the field at the end of 1858, and even then but few were coerced; they came in under the same conditions, and from first to last those conditions were a restoration to the status at annexation. No other terms were ever mentioned, and Sir Robert Montgomery can bear me out in the assertion that to reduce the Talookdars to submission a most unreserved settlement was made with them . . .

10. By fair arguments and politic treatment, then, the neck of the rebellion was broken in 1858 in Oudh. The main argument was the settlement as proprietor of his estate with each

Talookdar as he presented himself; not suddenly led on by some decisive blow, but man by man submission was tendered. As one gave in, another saw the treatment he received; so gradually was confidence restored; it was hardly a beaten foe claiming mercy at our hands, but men yielding to a political invitation, and relying, after his distrust was overcome, on the good faith of the British Government.

National Archives of India, No 30.

Question • In what sense was the relationship between the British and the Oudh Talukdars one of 'mutual interdependence'?

Source 5.8 The Pangkor engagement, 1874

Whereas, a state of anarchy exists in the Kingdom of Perak owing to the want of settled government in the Country, and no efficient power exists for the protection of the people and for securing to them the fruits of their industry, and

Whereas, large numbers of Chinese are employed and large sums of money invested in Tin mining in Perak by British subjects and others residing in Her Majesty's Possessions, and the said mines and property are not adequately protected, and piracy, murder and arson are rife in the said country, whereby British trade and interests greatly suffer, and the peace and good order of the neighbouring British Settlements are sometimes menaced, and

Whereas, certain Chiefs for the time being of the said Kingdom of Perak have stated their inability to cope with the present difficulties, and together with those interested in the industry of the country have requested assistance . . .

His Excellency Sir Andrew Clarke, K.C.M.G., C.B., Governor of the Colony of the Straits Settlements, in compliance with the said request, and with a view of assisting the said rulers and of effecting a permanent settlement of affairs in Perak, has proposed the following Articles of arrangement as mutually beneficial to the Independent Rulers of Perak, their subjects, the subjects of Her Majesty, and others residing in or trading with Perak, that is to say:

I. *First*. That the Rajah Muda Abdullah be recognised as the Sultan of Perak.

II. *Second*. That the Rajah Bandahara Ismail, now acting Sultan, be allowed to retain the title of Sultan Muda with a pension and a certain small Territory assigned to him . . .

VI. *Sixth*. That the Sultan receive and provide a suitable residence for a British Officer to be called Resident, who shall be accredited to his Court, and whose advice must be asked and acted upon on all questions other than those touching Malay Religion and Custom . . .

VII. *Seventh*. That the Governor of Larut shall have attached to him as Assistant Resident, a British Officer acting under the Resident of Perak, with similar power and subordinate only to the said Resident.

VIII. *Eighth*. That the cost of these Residents with their Establishments be determined by the Government of the Straits Settlements and be a first charge on the Revenues of Perak.

IX. *Ninth*. That a Civil list regulating the income to be received by the Sultan, by the Bandahara, by the Mantri, and by the other Officers be the next charge on the said Revenue.

X. *Tenth*. That the collection and control of all Revenues and the general administration of the Country be regulated under the advice of these Residents.

The above Articles having been severally read and explained to the undersigned who having understood the same, have severally agreed to and accepted them as binding on them and their Heirs and Successors.

C N Parkinson, pp. 323–5.

Question • Which do you think was the crucial clause of this agreement? Why?

Diplomatic Collaboration

Juggernaut large overpowering force or object; institution to which people blindly sacrifice themselves or others

Just as individual Asians had to become accustomed, whether they liked it or not, to living under European rule, so the governments of independent Asian states had to come to terms with the ever-expanding **Juggernaut** of Western imperialism, which had shown itself to be too powerful to be stopped by force. Some, like the Manchus of China, tried to buy time by giving the foreigners piecemeal concessions while simultaneously building up their military strength. This 'self-strengthening' approach will be discussed in the next chapter.

Other countries, however, took the view that it was better to seek the friendship of the Western Powers on the grounds that the West was not likely to attack countries that posed no threat to its interests. These countries included Japan, under the leadership of Emperor Meiji (*r.* 1867–1912), and Siam, under the rule of King Mongkut Rama IV (1804–68), who came to the throne in 1851, after spending twenty years as a Buddhist monk. In the words of Nguyen Truong To (1827–71), a Catholic, French-speaking mandarin, the key to independence was 'diplomacy'. His 1867 memorial to the throne on the subject of Vietnam's relations with the West is cited as source 5.10.

However, the rivalry that existed between the European Powers complicated matters: if an Asian country became too friendly with one Power, it was likely to incur the hostility of the others. The question which Mongkut had to answer was whether it was better for Siam to 'make friends with the crocodile' or 'hang on to the whale' (source 5.9).

Source 5.9 King Mongkut Rama on the choices facing Siam, 1867

Since we are now being constantly abused by the French because we will not allow ourselves to be placed under their domination like the Cambodians, it is for us to decide what we are going to do; whether to swim up-river to make friends with the crocodile or to swim out to sea and hang on to the whale. . . .

It is sufficient for us to keep ourselves within our house and home; it may be necessary for us to forego some of our former power and influence.

A L Moffat, p. 124.

Question • If France was the 'crocodile', who do you think was the 'whale'?

Source 5.10 Nguyen Truong To: Memorial on the utility of diplomatic relations, 1867

Because it is on an equal footing with Vietnam, we take the example of Siam. This country is neither larger nor stronger than ours. However, in her relationship with the Western powers, Siam had the wisdom to accept trade with England, Spain, and Portugal. Thereby the Siamese were able to preserve their independence. They are not obliged to defend their frontiers nor are they compelled to protect their interests. Furthermore, wherever they go, they are respected as if they were a world power. Although England and France are both eager to conquer Siam, neither of them dares decide to move forward. This very special position in the world Siam has obtained and maintained through her diplomacy; and the manner in which this diplomacy is conducted renders her richer and stronger day by day.

Under the present circumstance, the only efficient means of self-defense we have at our disposal is diplomacy conducted with a reasonable foreign policy. I see no better method for preserving our autonomy. Moreover, if we do not quickly decide on a style in our conduct of diplomacy and if we do not hasten to apply this style, I am afraid that when we finally do, it will be too late . . .

B L Truong, p. 91.

Question • Why did the 'Siam strategy' not work in the case of Vietnam?

Dependent Minorities

pecuniary
relating to money

As we have seen, **pecuniary** self-interest prompted some groups in colonial society, such as the aristocracy and the professional class, to collaborate with the foreigners. But the most ready-made collaborators, if one can put it that way, were indigenous racial or religious minority groups who, under the old regime, had suffered regular persecution for their customs and beliefs.

The Dyaks of Borneo warmed to the rule of the Dutch and the White Rajahs because, if nothing else, it brought to an end their persecution by the Malays. Similarly, the Karens, Mons and Shans welcomed British liberation from the tyranny of the Burmans. As the Japanese 'Mito School' writer Aizawa Seishisai (1782–1863) perceived, native converts to Christianity constituted a natural bastion of support for foreign rule (source 5.11).

Naturally, so did the Eurasians who, as well as carrying European blood in their veins, were almost all Christians. Eurasians were European in their tastes, manners and political allegiances—in everything, in fact, except the colour of their skin. They flourished under colonial rule, maintaining a number of key government jobs, especially in the railways (see source 5.4).

Yet this high profile was ultimately their undoing, for it caused them to be envied and despised by other natives. In source 5.12, novelist John Masters, who served in the Indian Army, vividly recreates for us the dilemmas that confronted the 'Anglo-Indians' on the eve of independence in 1947.

Source 5.11 Aizawa Seishisai: Christians as a fifth column, 1825

As to the Western barbarians who have dominated the seas for nearly three centuries—do they surpass others in intelligence and bravery? Does their benevolence and mercy overflow their own borders? Are their social institutions and administration of justice perfect in every detail? Or do they have supernatural powers enabling them to accomplish what other men cannot? Not so at all. All they have is Christianity to fall back upon in the prosecution of their schemes. . . . When those barbarians plan to subdue a country not their own, they start by opening commerce and watch for a sign of weakness. If an opportunity is presented, they will preach their alien religion to captivate the people's hearts. Once the people's allegiance has been shifted, they can be manipulated and nothing can be done to stop it. The people will be only too glad to die for the sake of the alien God.

W T de Bary (Ed), *Japanese Tradition*, p. 602.

Question • What events in Japanese history might explain Aizawa's assertion in the last sentence?

Source 5.12 The Eurasian dilemma

. . . when Mrs Jones didn't speak I took some more potatoes and said, 'Sir Meredith Sullivan is going to give us a talk during the whist drive.' I explained that the talk was going to be about St Thomas's, Gondwara—my old school.

When I mentioned St Thomas's, Victoria said thoughtfully, 'I suppose the Presidency Education Trust will have to close it. It's sad, though.'

I stared at her, and this thing that would have to come out got higher in me. I said, 'Close it? Close St Thomas's? They can't! Why, that was my old school.'

'I know, Patrick,' she said. 'You've got the tie on now.' I had. It has blue, yellow, and violet stripes. She said, 'But—things are going to change, aren't they? They *are* changing . . .' . . .

Now she was getting excited, and her eyes sparkled, and she didn't talk la-di-da, she talked the way we do. She said, 'I've been four years among only Englishmen and Indians. Do you realize that they hardly know there is such a thing as an Anglo-Indian community? Once I heard an old English colonel talking to an Indian—he was a young fellow, a financial adviser. The colonel said, "What are you going to do about the Anglo-Indians when we leave?" "*We*'re not going to do anything, Colonel," the Indian said. "Their fate is in their own hands. They've just got to look around and see where they are and who they are—after you've gone."' . . .

What the Trust was saying to us now, in 1946, was this: 'St Thomas's doesn't pay its way. It only survives because the Provincial Government gives it more help from provincial funds than it's really entitled to, considering the number of boys it educates. An Indian government will come to power soon. Is it likely that they will continue to give special help to the education of Anglo-Indians? Of course not! So sooner or later you'll have to sell out. But there's a boom on now, and *now* is the time to sell St Thomas's. With the money, you'll be in a position to keep the day-schools open, at least, whatever happens.'

All that made sense, I suppose, but what those Englishmen in Bombay didn't realize was that we *couldn't* sell St Thomas's, because it was in our hearts. It, the idea of it, was part of us. Without it we'd just be Wogs like everybody else. They might just as well have said we couldn't afford trousers or topis, or told us to turn our skins black instead of khaki . . .

Then Victoria shouted, ' . . . Do you think an Indian government, a Congress government, is going to keep on holding jobs open especially for us on the railways and telegraphs? Colonel McIntyre said—'

Then I shouted, 'Oh, to hell with Colonel McIntyre! I am sorry, Mrs Jones—but, Victoria, what does your Colonel McIntyre know about St Thomas's? Was he there? He was at Eton School, I bet!'

Victoria said, 'He knows nothing about St Thomas's! But he thinks the English will leave India very soon,' . . .

Then I banged my fist on the table. I shouted, 'My God, if they leave I will go Home with them.'

Victoria sat up with a jerk, very pale, and she screamed, 'Home? Where is your home, man? England? Then you fell into the Black Sea on your way out? I don't want to see the Congress ruling here, but I am only asking you, what else is there that can happen? I am only asking you to think, man!' She pushed back her chair and ran out of the dining-room.

I sat there, feeling a little sick. That last thing I had said, about going Home, was mere foolishness, and I knew it. The whole point that made it impossible to give way, even to argue, was that we *couldn't* go Home. We couldn't become English, because we were half Indian. We couldn't become Indian, because we were half English. We could only stay where we were and be what we were. Here Colonel McIntyre was right too. The English would go any time now and leave us to the Wogs.

J Masters, pp. 25–8.

Question • Is the author trying, in this passage, to enlist our sympathy for the Anglo-Indians of *Bhowani Junction*, or is he poking fun at them?

Beneficiaries of the Market Economy

In chapter 4 it was noted that many parts of south and South-East Asia under colonial rule achieved quite spectacular agricultural growth, especially in export-orientated 'cash'-crops. Those farmers who were prepared to take the risk of producing not only for themselves but for the market, by planting commercial crops, often became very rich. One such success story is described by American **anthropologist** Michael Adas in source 5.13.

anthropologist one who studies mankind, especially its societies and customs

Although it destroyed the traditional handicraft sector, Asia's exposure to the world market resulted in the creation of a number of new processing industries (for example sugar-refining) and eventually, a modern textile industry. At the same time, the colonial bureaucracies spawned many jobs in the service and construction sectors.

market economy economy in which crops are produced for sale rather than the grower's needs

Not everyone benefited from the **market economy**, however, and only a tiny few became rich. Source 5.14 comes from the memoirs of a Penang Chinese woman, 'Queeny' Chang, and shows clearly that those who prospered financially under colonialism were able to indulge in a lifestyle that previously would have been reserved for kings and nobles. It is hardly surprising that these people became, in general, strong supporters of foreign rule.

Source 5.13 From smallholder to landlord: Maung Kyaw Din of Burma

Until the first decade of the twentieth century, most landlords were (or at least began as) agriculturalists. Maung Kyaw Din . . . is a superb example of a small landholder who rose to become a large landlord. Maung migrated to the Maulamyainggyun area of the Myaungmya District from upper Burma in 1875. When he arrived, the tract was covered with jungle and, with the exception of a few families of fishermen, devoid of inhabitants. He and his family built a hut and after obtaining permission from the nearest subdivisional officer, they cleared and began to cultivate a small holding. In the first years Maung worked his land without draft animals, plow, or hired laborers. . . In order to earn extra cash, he sold firewood and bamboo on the side. Maung's paddy production rose gradually from 150 (forty-six pound) baskets in the first year to 1,000 baskets by the fifth year. At that time he sold 700 baskets for over Rs. 500 profit and bought a pair of bullocks. In the sixth year his holdings were greatly enlarged and produced 2,000 baskets. By the tenth year after his arrival his output reached 4,000 baskets, and he employed ten laborers and three yokes of cattle. During the land boom of the 1890s he lent money on a large scale to new settlers and in many cases added now holdings to his estate through foreclosure of mortgages given as security on loans which were not repaid.

By the first decade of the twentieth century Maung owned 750 acres, thirty draft animals which he rented out at a profit of Rs. 1000 annually, and a house worth Rs. 5,000. He had also financed the building of a *kyaung* (Buddhist monastery) and pagoda in his village, which cost an estimated Rs. 23 000. In the first years of the twentieth century Maung began to speculate in paddy. At the height of his affluence he handled some 5,000 baskets in a single season. Shortly thereafter he retired, leaving great wealth and extensive holdings to his six sons, who were also agriculturalists. Maung Kyaw Din was only one of thousands of diligent cultivators who achieved positions of wealth and local importance on the Delta frontier in the early phase of growth when land was readily available and the demand for and price of paddy rose almost continually.

M Adas, pp. 71–2.

Question • How much of Maung's success was due to British colonialism?

Source 5.14 A mansion in Penang, *c.* 1910

There were so many rich people with beautiful mansions in Penang. One family was as wealthy as the other, if not wealthier. Many of them had found their fortune in tin mining and they tried to outshine one another in the architecture of their grand homes. Each house was as luxuriously decorated as the other with gilded furniture, crystal chandeliers, Venetian glass lamps and bibelots.

One house which has remained in my memory is that of Chung Thye-phin. It was like a castle with two towers and was built on a hill. The garden stretched to the sea. Black marble steps led to a porticoed entrance with two rose-coloured marble Greek statues on either side. When we entered the big hall, my eyes were arrested by a life-size oil painting of an extremely handsome young man in a costume such as that worn by English lords: white breeches, sapphire blue long-tailed cutaway coat, frilly white shirt and a high cravat . . .

'That's the owner of this castle,' said Aunt Cheah Number Two, pointing to the picture.

At this moment, a beautiful young woman came down the stairs. She was very fair, her face a perfect oval with a straight nose and full red lips. Her eyes were bright and smiling. She was wearing a pale green gauze *kebaya* with a large floral design and a brown batik *sarong*. Her black hair was done in a style like my mother's with jasmines to adorn it and she wore exquisite jewellery.

'She has the most beautiful jewellery in all of Penang,' whispered Aunt Cheah again. Turning to the young woman who had reached the foot of the stairs to greet us, she said aloud, 'I've brought my relative from Medan to make your acquaintance and to see your house.'

'You're very kind,' Mrs Chung acknowledged in a sweet voice, 'you are always welcome.'

She led us through the house from one room to another, each laid out differently. Then she took us upstairs to their living quarters.

Her bedroom was separated from that of her husband's by a cosy drawing room and a study . . .

The bedroom was an absolute dream. The walls were panelled with *bois-de-rose* [rosewood] brocade which was also used for draperies for the windows over cream-coloured lace curtains. The soft sunlight shimmered through accompanied by a cool breeze from the sea. The bed was in the form of a big shell covered with a canopy decorated with silver cupids. The pillows and bedspread were of pink lace and satin. On the floor lay a fleecy cream-coloured carpet and the high ceiling was painted with lilies-of-the-valley and forget-me-nots . . .

A connecting door opened to the bathroom laid with rose tiles. Even in this room everything was luxurious: sets of towels in matching colours, soaps, bathcubes and other curiosities. Seeing these, Aunt Cheah remarked: 'If I had to use all these, I'd rather go without a bath.' It made us all laugh, but I felt for the first time something like envy. To be surrounded by so much luxury and beautiful things all the time must be marvellous, I thought.

Finally, Mrs Chung took us downstairs to have tea in the dining room built under the sea. We sat at a long table laid with all sorts of delicacies. We really enjoyed the tea and I ate to my heart's content. When I happened to look up at the ceiling I saw that it was not painted as I had at first thought. It was a glass dome through which I could see fishes swimming about!

Seeing my astonishment, Mrs Chung explained amiably: 'Yes, they are real fishes. My husband designed this room himself and had it built under the sea. He claims that it will relieve our boredom if we have our meals in the company of fishes. It is a pity that he is now in London buying racehorses but he will be back for the Gold Cup next month. Then, he can show you more of his eccentricities.'

Q Chang, pp. 51–4.

Question

- How might Mr Chung have acquired the money to build his palatial mansion?

The Nazim of Hyderabad's residence, Ootacamund, India. The Nazim modelled the building on a chalet he saw in Switzerland.

Beneficiaries of Modernisation

Asians welcomed Western colonial rule also for the technological marvels that it brought into their lives: steam and electricity, printed books, wonder medicines, photographs, and air-travel. On one level, these symbols of modernity amazed and excited. 'Minke', the Javanese high school student who is the narrator-hero of Toer's banned Indonesian novel, opens his story of growing up under Dutch rule by confessing to being addicted to all things modern. As a consequence of this addiction to modernity, he becomes increasingly fed up with his own narrow existence (source 5.15).

Unlike Minke, the ordinary people did not generally have access to the benefits of a Western education; for them the appeal of modernisation was more mundane, down-to-earth. Quite simply, it made life more comfortable. As Punjabi writer Ved Mehta wrote in his family biography, even something as basic as piped water was a miracle in a society which had always had to depend on wells and rivers (source 5.16).

Revelling in the delights of fast train travel and kerosene lamps and fresh running water, Asians could not help but feel grateful to those Westerners who, in Minke's words, had 'worked so tirelessly to give birth to these new wonders'.

Source 5.15 'Minke' on the marvels of Western science

One of the products of science at which I never stopped marvelling was printing, especially zyncography. Imagine, people could reproduce tens of thousands of copies of any photo in just one day. Pictures of landscapes, big and important people, new machines, American skyscrapers, everything and from all over the world—now I could see them for myself upon these printed sheets of paper. How deprived was the generation before me—a generation that was satisfied with the accumulation of its own footsteps in the lanes of its villages. I was truly grateful to every single person and indeed to all people who have worked so tirelessly to give birth to these new wonders. Five years ago there were still no printed pictures, only block and lithographic prints, which could not yet represent reality as it truly was.

Reports from Europe and America brought word of the latest discoveries. Their awesomeness rivalled the magical powers of the gods and knights, my ancestors in the *wayang* [puppet play]. Trains—carriages without horses, without cattle, without buffalo—had been witnessed now for over ten years by my countrymen. And astonishment still remains in their hearts even today. Batavia–Surabaya can be traversed in only three days! And they're predicting it will be only a day and a night! A day and night! A long train of carriages as big as houses, full of goods, and people too, pulled by water power alone! If I had ever, during my life, met Stephenson, I would have made an offering to him of a wreath of flowers, all of orchids. A network of railway tracks splintered my island, Java. The trains billowing smoke coloured the sky of my homeland with black lines which faded and faded into nothingness. It was as if the world no longer knew distance—it had been done away with too by the telegram. Power was no longer the monopoly of the elephant and the rhinoceros. They had been replaced by small man-made things: nuts, screws and bolts.

And over there in Europe, people had begun making smaller machines still, with even greater power, or at least with the same power as steam-engines. Indeed not with steam. With oil. Vague reports were saying: a German had even made a vehicle which is moved by electricity. *Ya Allah* [Oh God!], and I still didn't know how to prove what that thing electricity was.

The forces of nature were beginning to be changed by man so as to be put to his service. People were even planning to fly like *Gatotkaca* [a *wayang* character], like Icarus. One of my teachers had said: a little while, just a little while, and humankind will no longer have to force their bones and squeeze out their sweat for so little result. Machines would replace all and every kind of work. People will have nothing to do except enjoy themselves. You are fortunate indeed, my students, he said, to be able to witness the beginning of the modern era here in the Indies.

Modern! How quickly that word had surged forward and multiplied itself like bacteria throughout the world. (At least, that is what people were saying). So allow me also to use this word, even though I still can't truly fathom its meaning.

P Toer, pp. 2–4.

Question

- This extract (like several others in the book) comes from a novel. Under what circumstances might a work of fiction be used legitimately as a historical source?

Source 5.16 Piped water: A modern miracle in Lahore, *c.* 1905

Behind the bazaars and lanes were rabbit warrens of *mohallas*, or blocks of tenement houses, opening onto squares that were entered by still narrower *gullis*, and these *mohallas* and *gullis* contained more life and variety than could be found in any village. Bhaji Ganga Ram, however, had taken lodgings in a new, open part of the city, where the British government officers and the Indians attached to the British establishment lived. These lodgings—three rooms and a kitchen, one flight up—were somewhat better than the ones in Multan. But what Daddyji liked most of all about them was a discovery he made within minutes of reaching them. Just outside the building was a curved pipe sticking up out of the ground, with a handle on top. When the handle was raised and lowered, water came gushing out of the mouth of the pipe, which was high enough to sit under. Daddyji had never seen anything like it, and the first thing he did in Lahore was to take a bath.

V Mehta, pp. 35–6.

Question

- How would 'Daddyji' have bathed in Multan or in the village where he was born.

A street letter-writer in Rawalpindi, Pakistan. In a largely illiterate society, letter-writers have always been needed, but where they once used a pen, they are now assisted by modern technology.

6 *The Cult of Westernisation*

Alas, let us moderate our love for
the fatherland;
The Wind from the West has blown
lightly over us and made us
shiver and tremble.

Michel Duc Chaigneau, 1875

ASIAN COUNTRIES UNDER DIRECT colonial rule could hardly avoid being Westernised to some extent. But what was the appeal of 'modernisation' to places such as Siam, China, Korea, and Japan which remained outside the umbrella of European colonialism? This chapter looks at the motives of Asian reformers. It argues that Western techniques and ideas were initially adopted in these countries for reasons of defence, but that modernisation gradually adopted a momentum of its own, generating changes far greater than its original sponsors expected or intended. Along with chapter 5, the experience of the natives living under European colonial rule is emphasised.

KEY DATES

1815 Bengali Brahmin scholar and reformer Raja Rammohun Roy (1772–1833) writes *The Precepts of Jesus.*

1845 The first four Asians to study abroad arrive in London from Bengal.

1853 Commodore Matthew Perry (1794–1858) of the USA and his squadron of 'black ships' sail uninvited into Uraga Bay, Japan.

1857 Institute for the Investigation of Barbarian Books founded in Edo (Tokyo).

1863 Foreign Language School established in Shanghai by reforming Chinese Viceroy Li Hung-chang (1823–1901).

1868 Tokugawa Shogunate overthrown by Western clans in the 'Meiji Restoration', which restored Japan to direct monarchical rule under Emperor Meiji Tenno.

1871 Samuel Smiles's *Self Help*, very widely read in Victorian England, is translated into Japanese. It becomes one of the most popular books of the Meiji era, selling one million copies.

1896 Playing cricket for Cambridge University, Prince Ranjitsinhji (1872–1933), afterwards Maharaja of Nawanagar, scores 2780 runs to beat W G Grace's longstanding county record.

1897 Chinese scholar-official K'ang Yu-wei (1858–1929) publishes *Confucious as a Reformer.*

1913 (Sir) Rabindranath Tagore (1861–1941) awarded the Nobel Prize for his poem *Gitanjali.*

1920 Westernising 'Young Turk' movement led by war-hero Mustapha Kemal 'Ataturk' (1881–1938) seizes power in Constantinople.

1925 Men and women dance together for the first time in Turkey.

1970 Japanese author Yukio Mishima (b.1925) commits *sepuku* (ritual suicide) on the balcony of Eastern Army headquarters in Tokyo in protest against the erosion of Japan's martial traditions.

ISSUES AND PROBLEMS

From the beginning of their contact until the end of the 18th century, the cultural traffic between Europe and Asia was mostly from East to West. Tea-drinking, Japanese furniture, the habit of bathing, pyjamas, one-storied houses in the Bengal style, or *bungalows*, Buddhist and Confucian philosophy (Voltaire thought the Chinese system of government the finest that had ever been created), polo, Indian-made silks and *muslins*, porcelain or 'china', as it became known in Europe, curry, and rhubarb were just a few of the products, customs and ideas taken up by Westerners for their own use during the first phase of the imperial age. Since the end of the 18th century, however, the tide has been nearly all the other way. Slowly, at first, but then with increasing rapidity, Asia was remodelled in the image of the West.

It is not surprising that this happened in the areas under direct colonial rule, since one of the major goals of colonial policy was to reform and regenerate native society. But what of the countries that remained at least nominally independent, such as Siam, Persia, Japan and Turkey? Not only did all these places modernise during the 19th and 20th centuries by borrowing heavily from Europe and the United States, but the last two, at least, underwent far more sweeping cultural transformations than any of the Asian colonies. By the Meiji Restoration of 1868, which is usually taken as the starting point of Japan's transformation, the Tokyo government had financed seven overseas fact-finding missions. Over the next forty years it brought 2400 foreigners to Japan to teach Western skills. At the same time it embarked on a massive railway building program; by 1910 6000 miles of track had been laid. In *Memories of a Meiji Boyhood* Bunroku Shisi recalls that in Tokyo where he lived, the 'Nights became bright with gas and electric illumination . . . [while] Diseases long regarded as incurable surrendered to improved medical science.'[1]

Half a century later, Turkey underwent a similar renaissance. When the charismatic war-hero Mustapha Kemal seized power in 1922, he embarked on a roller-coaster reform spree designed, as one of his followers put it, to 'blow up' the 'bridges joining Turkey to the Middle Ages'.[2] In the space of ten years Kemal abolished the monarchy, romanised the alphabet, closed the Islamic courts, drafted a new constitution which set Turkey on the road to democracy, ordered women out of **purdah**, and introduced ballroom dancing. What prompted these states to Westernise? Were they coerced in some fashion? Or did they embrace modernisation freely and willingly in the hope that it would lead to the creation of a better life for their citizens?

purdah seclusion of women in Muslim and Hindu society

Another puzzling aspect is the popular Asian response to the West. Particularly in East Asia, foreigners were looked down upon. Well into the 19th century the colloquial epithet for Westerners was 'barbarians'—a term which, in some cases, may have been well deserved but which, nevertheless, speaks of arrogance and racism. Moreover this attitude was not confined to the common people, being shared even by educated and curious men like Japan's Hirata Atsutane (source 6.1). Yet, very soon, intellectuals and peasants alike were eagerly partaking of the good things of the West. For example, railways everywhere were popular immediately with the people. Indian railways—the biggest system—travelled 14.7 billion passenger-miles in 1912. In 1900 Tokyo Station serviced more passengers than New York's Grand Central. Similarly, Asians flocked to acquire Western education. In 1845 the first Indian students arrived in London and twenty years later the first Vietnamese students disembarked at Marseilles. By the end of the century, some 30 000 Chinese, Japanese and Indians had studied abroad. Finally, some Asians (men especially) developed a taste for the European life-style, consuming meat and alcohol, and forsaking their *sarongs* and *dhotis* for trousers, frock-coats, and shirts with starched collars. What caused this dramatic turnabout in attitude, from contempt to envy and imitation?

There is also the question of the social impact of Westernisation to be considered. Although modernisation is usually thought of as a 'good thing', too rapid wholesale change can cause dislocation and trauma. Throw out the past, and you destroy your roots. Alert to this danger, the Westernising class tried to avoid the pain by borrowing selectively and attempting to **syncretise** Western and Eastern ideas. However, as the Chinese, in particular, discovered, it was not all that easy to separate Western science and Western culture. How much, in the end, did the modernisers salvage of Asia's traditional values?

syncretise deliberate merging of two different philosophies or religions

THE DEBATE

Recent writing on the subject of Westernisation in Asia has focused on two main issues: first, the motives of the Westernisers, and second, the impact of Westernisation on Asian society. Opinion on the first question is divided into those who think that the Westernising push was a means to an end, namely national **self-strengthening**, and those who believe that it reflected a genuine commitment to European 'enlightenment' values such as equality, democracy, and the rule of law. On the second question, there is a fundamental disagreement between those who, following Karl Marx (source 9.8), hold that European colonialism performed a thorough, and on the whole salutary, clean-up job in Asia, and those who think that the Western impact was only of marginal significance. Representative of the latter group is American geographer Rhoads Murphey. After an exhaustive analysis of the colonial port cities of China, Murphey concluded that:

self-strengthening modernisation by countries so as to resist further Western advances

> For all its vigor and self-confidence, the foreign presence and its efforts at innovation made only the smallest of dents in the material fabric of traditional China. The great majority of Chinese were unaffected, directly or indirectly, in their way of life.[3]

The problem here, of course, is the limited nature of the evidence. Even if Murphey were right about China, we cannot assume that other parts of Asia were equally untouched.

Initial Reactions to the 'Barbarians'

Despite centuries of often receptive contact—the Jesuits Matteo Ricci and Adam Schall won favour and high rank in Ming China, and for a century (1540–1640) Japan opened its doors wide to Christianity—the dominant attitude to foreigners in pre-modern east Asia was one of arrogance and suspicion. In 1614, the Shogun Toyotomi Hideyoshi expelled all Christian missionaries from Japan, and in 1641 the Edo government adopted a policy of *sakoku* (isolation). The country was closed off from all contact with foreigners except for the Chinese and the Dutch, who were allowed to send one ship a year from Batavia and to maintain a factory (the Dutch factors thought it more like a prison) on the manmade island of Deshima in Nagasaki harbour.

Similarly, in 1760 the Ch'ing government in Peking imposed the so-called 'Canton System', limiting foreign contact to the southern port of Canton and subjecting those who ventured to trade there to a humiliating set of restrictions. This policy was still in force at the end of the 18th century (source 2.10).

And in Vietnam, the Hue Mandarinate abandoned its previous policy of discouraging Christianity and instituted a program of utter persecution, culminating in the edict of 1833 (source 6.13).

It was not that Asians were not curious about what lay overseas. Scholars, especially, were very curious, and in Japan a whole school of foreign studies (*rangaku*) developed during the late 18th century. The Meiji reformer Fukuzawa Yukichi (1834–1901) notes in his autobiography (source 6.3) that this interest in things Western had to be kept secret for fear of reprisals from the people, even as late as the mid-19th century.

The people generally shared the government's dislike of foreign intrusion, as shown by the description of the incident at Uraga written by Mito School writer Sakuma Shozun (1811–64) (source 6.2). Indeed, even some Rangaku scholars were prejudiced. Hirata Atsutane (1776–1843) studied the Dutch and other Europeans and wrote much about them, but he could never hide his contempt for them (source 6.1). Who would have thought that the Japanese would soon become Asia's most aggressive Westernisers?

Source 6.1 Hirata Atsutane on the physical peculiarities of Europeans, *c.* 1810

As everybody knows who has seen one, the Dutch are taller than other people and have fair complexions, big noses, and white stars in their eyes. By nature they are lighthearted and often laugh. They are seldom angry, a fact that does not accord with their appearance and is a seeming sign of weakness. They shave their beards, cut their nails, and are not dirty like the Chinese. Their clothing is extremely beautiful and ornmented with gold and silver. Their eyes are really just like those of a dog. They are long from the waist downwards, and the slenderness of their legs also makes them resemble animals. When they urinate they lift one leg, the way dogs do. Moreover, apparently because the backs of their feet do not reach to the ground, they fasten wooden heels to their shoes, which makes them look all the more like dogs. This may explain also why a Dutchman's penis appears to be cut short at the end, just a like a dog's. Though this may sound like a joke, it is quite true, not only of Dutchmen but Russians. Kōdayū, a ship's captain from Shirako in Ise, who some years ago visited Russia, recorded in the account of his travels that when he saw Russians in a bathhouse, the end was cut short, just like a dog's. . . . This may be the reason the Dutch are as lascivious as dogs and spend their entire nights at erotic practices. . . . Because they are thus addicted to sexual excesses and to drink, none of them lives very long. For a Dutchman to reach fifty is as rare as for a Japanese to live to to be a hundred. However, the Dutch are a nation given to a deep study of things and to fundamental investigations of every description. That is why they are certainly the most skilled people in the world in fine works of all sorts, and excel in medicine as well as in astronomy and geography.

D Keene, p. 170.

Question • Would you describe this document as 'racist'?

Source 6.2 A Japanese takes revenge on Commodore Perry, 1853

Last summer the American barbarians arrived in the Bay of Uraga with four warships, bearing their president's message. Their deportment and manner of expression were exceedingly arrogant, and the resulting insult to our national dignity was not small. Those who heard could but gnash their teeth. A certain person on guard in Uraga suffered this insult in silence, and, having been ultimately unable to do anything about it, after the barbarians had retired, he drew his knife and slashed to bits a portrait of their leader, which they had left as a gift. Thus he gave vent to his rage.

W T de Bary (Ed), *Japanese Tradition*, p. 614.

Question • Was this a sensible response?

Source 6.3 Fukuzawa Yukichi on the risks of being a student of Western learning

. . . in the beginning, people simply hated the foreigners because all foreigners were 'impure' men who should not be permitted to tread the sacred soil of Japan. Among these haters of foreigners, the samurai were the most daring and having their two swords conveniently at their sides, some of the younger and less restrained of these would spring on the 'red-haired outlanders' in the dark. Still there was no reason for them to turn on the subjects of Japan, and so the students of foreign culture were yet safe from attack. While studying in

Osaka, and even after coming to Yedo to teach, I had no feeling of danger for several years. For instance, when I heard of an attack on a Russian in Yokohama soon after the opening of the port, I was merely surprised by the cruel incident, but I felt no personal concern about it.

Very quickly, however, the hatred of foreigners went through a tremendous development. It became more systematized, the objectives came to include many more persons, and the methods of slaughter became more refined. Moreover, political design was added to it and since the assassination of Chancellor Li in 1860, the world seemed to become tense with bloody premonitions in the air.

Tezuka Ritsuzo and Tojo Reizo were attacked by the Choshu clansmen for the simple reason that they were scholars of foreign affairs. Hanawa Jiro, a scholar of national literature, had his head cut off by an unknown man because of his sympathy for foreign culture. And the stores dealing in foreign goods were attacked for no other reason than that they sold foreign commodities which 'caused loss' to the country.

Here then was the beginning of the national movement. 'Honor the Emperor and Expel the Foreigners.' It was claimed that the Shogun was not prompt enough in carrying out the desires of the imperial court which had decreed the expulsion of all foreigners without exception. From this, it was argued that the Shogun was disobedient, was disrespecting the great doctrine of the land, and moreover was catering to foreign aggressiveness. Following this train of argument, it was but a step to calling all scholars of foreign culture traitors. And now we had to be careful. Especially when I heard of the attack on my friends and colleagues, Tojo and Tezuka, I knew that the hands of the assassins were not far from my door. Actually I was to go through some very narrow escapes.

The period from the Bunkyu era to the sixth or seventh year of Meiji—some twelve or thirteen years—was for me the most dangerous. I never ventured out of my house in the evenings during that period. When obliged to travel, I went under an assumed name, not daring to put my real name even on my baggage. I seemed continually like a man eloping under cover or a thief escaping detection.

Y Fukuzawa, pp. 226–8.

Question • How would Fukuzawa have replied to the accusation that he and other devotees of Western learning were 'traitors'?

Self-Strengthening by Asian Countries

xenophobia irrational hatred of foreigners

One of the things which caused East Asian governments to lessen their **xenophobia** was the gradual realisation that they could not keep the Europeans out even if they wished to. This realisation was brought about by bitter defeat and humiliation in military encounters such as that of 1839–42 in China and 1863 in Japan. Reluctantly acknowledging that the West was militarily invincible, for the time being, Asian-rulers swallowed their pride and started to rebuild their armies and navies on modern, Western lines. They hoped that such reforms would allow them, in the not too distant future, to reclaim their land.

Yet this transformation in attitude did not take place overnight, or without much soul-searching. In Japan, the process of readjustment was fairly quick. Begun soon after the unexpected appearance of Commodore Matthew Perry's 'black ships' in Uraga Bay in 1853, it was increased after the clash between Choshu forces and a British naval squadron in 1863. Readjustment became national policy after the overthrow of the Tokugawa Shogunate by the western clans in 1868. As source 6.5 shows, the leaders of Meiji in Japan were committed, at the highest level, to seeking knowledge 'throughout the world' in order to 'strengthen the foundations' of the State.

In China, however, Confucian bureaucrats resisted change. Brought up to believe that the 'Middle Kingdom' was superior in all things, they slowed the pace of modernisation to a crawl, much to the despair of forward-looking and pragmatic officials. The Governor-General of Chihli Province, Li Hung-Chang (1823–1901), was one such official frustrated by the resistance, and his 1872 memorial to the throne, enlisting the Emperor's sympathy, is reproduced as source 6.4.

The Iwakura mission to the USA and Europe sets out from Yokohama in 1871.

Why did Japan adapt so much more quickly? One factor was its size. Living on a crowded offshore island that was poorly endowed with natural resources, the Japanese had much more to gain by industrialisation which, as the example of Britain showed, could turn even a small country into a Great Power. Another factor was Japan's history of borrowing from China, which probably made it more receptive to foreign ideas. Third, it is possible that the Japanese collectively were driven to Westernisation by a deep-seated desire for international recognition and approval. Journalist Endymion Wilkinson surveys this Freudian idea in source 6.6.

Source 6.4 Li Hung-Chang on the 'method of self-strengthening', 1872

The Westerners particularly rely upon the excellence and efficacy of their guns, cannon, and steamships, and so they can overrun China. The bow and spear, small guns, and native-made cannon which have hitherto been used by China cannot resist their rifles, which have their bullets fed from the rear opening. The sailing boats, rowboats, and the gunboats which have been hitherto employed cannot oppose their steam-engine warships. Therefore, we are controlled by the Westerners.

To live today and still say 'reject the barbarians' and 'drive them out of our territory' is certainly superficial and absurd talk. Even though we wish to preserve the peace and to protect our territory, we cannot preserve and protect them unless we have the right weapons. They are daily producing their weapons to strive with us for supremacy and victory, pitting their superior techniques against our inadequacies, to wrangle with and to affront us. Then how can we get along for one day without weapons and techniques?

The method of self-strengthening lies in learning what they can do, and in taking over what they rely upon. Moreover, their possession of guns, cannon, and steamships began only within the last hundred years or so, and their progress has been so fast that their influence has spread into China. If we can really and thoroughly understand their methods—and the more we learn, the more improve—and promote them further and further, can we not expect that after a century or so we can reject the barbarians and stand on our own feet?

S-Y Teng and J K Fairbank, p. 109.

Question • What differences (if any) can you detect between Li Hung-Chang's position and that of the Meiji rulers of Japan (source 6.5)?

Source 6.5 The Meiji charter oath, 1868

By this oath we set up as our aim the establishment of the national weal on a broad basis and the framing of a constitution and laws.

1. Deliberative assemblies shall be widely established and all matters decided by public discussion.
2. All classes, high and low, shall unite in vigorously carrying out the administration of affairs of state.
3. The common people, no less than the civil and military officials, shall each be allowed to pursue his own calling so that there may be no discontent.
4. Evil customs of the past shall be broken off and everything based upon the just laws of Nature.
5. Knowledge shall be sought throughout the world so as to strengthen the foundations of imperial rule.

W T de Bary (Ed), *Japanese Tradition*, p. 604.

Question • Why was this seen at the time as an extremely radical program?

Source 6.6 A psychological theory of Japanese behaviour

There is a persuasive interpretation of the roots of Japanese individual behaviour by the psychiatrist Doi Takeo that holds that the fundamental characteristic of the Japanese is his desire, his need, to be loved (*amae*) by an understanding and respected parental or big brother figure on whom he can depend. To extend Professor Doi's concept of the role of *amae* to Japan as a nation: in the past Japan felt not only unique but also isolated. In looking to China, to Europe and then to America as models she derived a sense of belonging. If Japan acted in consonance with the patterns laid down by these models she would win their approval and affection. Conversely the parental model figure was expected to play its role in extending understanding to Japan. Although there are signs, as we shall see, that Japan is beginning to break out of this role playing, nevertheless, even to this day, the pattern persists.

E Wilkinson, p. 91.

Question • Did other nations suffer from the 'need to be loved', or was it something unique to Japan?

The New Mecca

In their struggle to come to terms with the new European-dominated international order, the Chinese clung doggedly to the belief that their culture was superior and did not need to be radically altered because of the Western threat. This approach was encapsulated in the slogan, *Chung-hsueh wei t'i, Hsi-hsueh wei yung*, which, roughly translated means 'Chinese learning for the fundamental principles, Western learning for the applications'.

However, it was difficult to put this delicate distinction into practice: in order to manufacture modern guns the Chinese first had to study chemistry; in order to use them effectively on the battlefield they had to acquire a knowledge of mathematics. This sparked a broader interest in Western science, and from there it was but a short step to European philosophy and economics. By this circuitous route, intellectuals like the 'Hundred Days' reformer T'an Ssu-t'ung gradually came to appreciate that the West had much more to offer than ships and guns (source 6.7).

Again, the self-strengthening program necessitated the hiring of large numbers of European and American advisers and technicians, some of whom came in time to occupy key administrative positions—especially in Siam (source 6.9). Although strenuous efforts were made to keep these 'hired hands' socially isolated, their presence led inevitably to a mutual exchange of ideas.

Similarly, Asian leaders were drawn irrevocably into closer contact with Europeans by the awkward but necessary business of opening foreign relations with the Western Powers.[4] As we have seen, King Mongkut Rama recognised that Siam's survival as an independent country hinged on keeping on good terms with the British and playing them off against the French. As his letter welcoming the first British ambassador Sir John Bowring (source 6.8) suggests, however, his eagerness to see more of the British stemmed mainly from a wish to know more about their civilisation

Source 6.7 T'an Ssu-t'ung on the need for complete Westernisation in China, 1898

Your letter says that during the last several decades Chinese scholars and officials have been trying to talk about 'foreign matters' (*yang-wu*), but that they have achieved absolutely nothing and, on the contrary, they have been driving the men of ability in the empire into foolishness, greed, and cheating. Ssu-t'ung thinks that not only do you not know what is meant by 'foreign matters,' but also that you are ignorant of the meaning of discussion. In China, during the last several decades, where have we had genuine understanding of foreign culture? When have we had scholars or officials who could discuss them? If they had been able to discuss foreign matters, there would have been no such incident as we have today [the defeat of China by Japan]. What you mean by foreign matters are things you have seen, such as steamships, telegraph lines, trains, guns, cannon, torpedoes, and machines for weaving and for metallurgy; that's all. You have never dreamed of or seen the beauty and perfection of Western legal systems and political institutions. . . All that you speak of are the branches and foliage of foreign matters, not the root. . .

We should extend the telegraph lines, establish post offices to take charge of postal administration, supply water, and burn electric or gas lamps for the use of the people. When the streets are well kept, the sources of pestilence will be cut off, when hospitals are numerous, the medical treatment will be excellent. We should have parks for public recreation and health. We should have a holiday once every seven days to enable civil and military officials to follow the policy of (alternation between) pressure and relaxation. We should thoroughly learn the written and spoken languages of all countries so as to translate Western books and newspapers, in order to know what other countries are doing all around us, and also to train men of ability as diplomats. We should send people to travel to all countries in order to enlarge their points of view and enrich their store of information, to observe the strengths and weaknesses, the rise and fall, of other countries; to adopt all the good points of other nations and to avoid their bad points from the start. As a result there will be none of the ships and weapons of any nation which we shall not be able to make, and none of the machines or implements which we shall not be able to improve. We should be exact about our units of measure, examine our legal system, and unify out moral standards and customs. When our legal system is established, our culture will be kept intact. . .

Your idea of despising our enemies arises because you think that they are still barbarians. This is a common mistake of the scholars and officials of the whole empire and they must get rid of it. A proverb says, 'Know yourself and know your enemy' ['and in a hundred battles win a hundred victories']. We must first make ourselves respectable before we despise others. Now there is not a single one of the Chinese people's sentiments, customs, or political and legal institutions which can be favorably compared with those of the barbarians. Is there any bit of Western culture which was influenced by China? Even if we beg to be on an equal footing with the barbarians, we still cannot achieve it, so how can we convert them to be Chinese? . . .

S-Y Teng and J K Fairbank, pp. 158–60.

Question • How does T'an's attitude to the West differ from that of the earlier self-strengtheners (source 6.4)?

No 37.

Rajmondirn house
grand palace
Bangkok.
27th March 1855.

My gracious friend

It give me today most rejoyful pleasure to learn your Excellency's arrival here as certainedly as your Excellency remained now on board the Steamer "Rattler" which accompanied with a brig of war. I can not hesitate to send my gladful cordil more than an hour. I beg to send my private minister Mr "Nai Kham Nai Phrong & Mr 'Nai Bhoo" with Som Siamese fruits for showing

To His Excellency Sir John Bowring
Right Dr of Laws &c &c &c

. letter in King
longkut Rama's own
andwriting addressed to
is 'gracious friend', the
ritish ambassador.

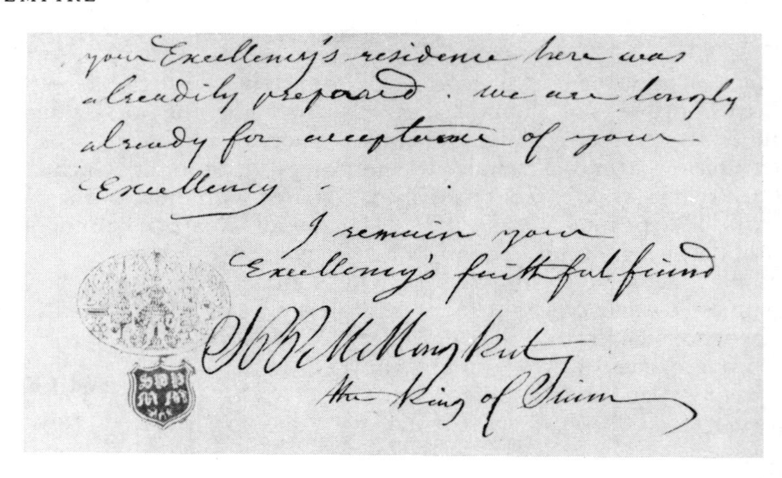

your Excellency's residence here was already prepared. we are longly already for acceptance of your Excellency.

I remain your Excellency's faithful friend

S P P M Mongkut
the King of Siam

Source 6.8 King Mongkut Rama display his prowess in English, 1855

Rajmondern House,
Grand Palace, Bangkok,
23rd April, 1855.

To-day morning I have forgot to ask your Excellency for my certain necessity; permit to pursue your Excellency with this note.

Can your Excellency give me two or three shels which for cannon or mortar on board *Rattler*, and two or three *rockets*, if there were on board, which (the shells and rockets) are newly improved? I wish to observe and consider and refer to our workmen whether would they be able to imitate or follow such or similar mode thereof for our use here. But my mind will be not hurt or troubled entirely, if your Excellency, on seeing of any object, would say that they are not allowable.

Also before this time once I have seen a model of an article I think now needable for may own use. Allow me to say for the said thing.

There is a chair like an arm-chair; in forehead of its arms there is handle erected holding the book and magnifying-glass for the reader seating upon the chaire: it called reading chair or seat. I wish to follow the model; but our workmen could not do, for some things therein are not exact in the picture.

Will your Excellency order from England to bring me one or two of such reading chairs?

I think it would fit for old man to be happy on reading of the English books.

N.B.—Whatever I have ordered to your Excellency, I shall pay their stated cost fully according to your Excellency's faithful statement in their price, when the articles were arrived with stated cost.

Please favour me by declaration for pardon me in writing to her Britanic Majesty.

The translation of Siamese letter written in English by myself, translated by myself alone; but it was written firstly in draft, which I have revised in every one word with English Dictionary: so the letter of which the copy I have given your Excellency can be said right considerably according to Siamese idiom and style.

But I am very afraid for my second letter, written in English at once in the paper sent in box with draf: it was not revised again, when I have written hurrily throughout the day break or broken. I have but a copy pressed in my letters-book, from which I will take copy and send your Excellency in pursuence to China *viâ* Singapore, with catalogue of the Royal presents. But meaning of my letter I trust can be understood by English readers every one word—some fine, some would be blameable.

This from your very faithful friend,

(Signed) S. P. P. M. MONGKUT
Rex Siamensium.

J Bowring, pp. 436–8.

Question • What do you think Sir John Bowring would have made of this letter?

Source 6.9 Foreigners in the service of the Siam Government, 1909

eneral Advisers, in foreign affairs, finance, agriculture, etc.	6
eneral financial agent of the government	1
esser advisers, in education, etc.	6
egal advisers, probationary legal advisers, and assistants	21
irector-generals of departments or equivalent	13
ssistant director-generals or equivalent	23
oreigners engaged in administrative work at the level immediately below department management, including various inspectors	69
rchitects and engineers, civil, mechanical, etc., not otherwise classified	51
ther engineering-type technicians	40
ducators not otherwise classified	14
awyers not otherwise classified	12
aval ship captains	4
ssistants to naval ship captains	2
aval engineering officers	2
redge masters	4
arbor vessel captain	1
edical doctors not otherwise classified	6
riculture specialists (Japanese)	3
atisticians	2
ccountants not otherwise classified	2
hemists not otherwise classified	2
urse	1
eterinarian	1
terpreter	1
ocomotive engineers	15
mbassy councilors	3
mbassy secretaries	5
mbassy attaché	1
onsuls general	8
otal	319

W J Siffin, p. 97.

Question

- Did the presence of so many foreigners in the administration undermine Siam's independence?

Importance of a Western Education

The previous section explained how, almost despite themselves, Asian countries were persuaded to adopt the trappings of Western military hardware. Then, slowly, imperceptibly, other products of the Western mind were absorbed into Asian culture.

However, decisions about the proper pace and extent of Westernisation were not by any means the exclusive province of governments. All over Asia during the later part of the 19th century and into the 20th century, ordinary men and women were facing similar choices in their own lives. Should they send their son to a European school? Should they—could they—invite a

A Japanese couple in the Meiji Era (1868–1912). The husband and wife have adopted the latest in Western fashion. To modern eyes, they look rather absurd. Yet is the situation so different today?

foreigner into their home? Would wearing Western dress enhance their prospects at work? Could one be 'modern' while holding fast to caste restrictions on marriage, dining and overseas travel? The answer, for most Asians, was a definite 'yes'. But why? What was it that made Western learning and life-styles attractive?

Perhaps their main attraction, not surprisingly, was that they held the key to success. In the colonies, particularly, government employment was, as we have seen, much sought after. The prerequisite for such employment was a Western education, as an anonymous poem published in the *Ceylon Observer* newspaper in 1848 wryly pointed out (source 6.10).

But that is not all. As well as widening career opportunities, the acquisition of a good Western education could also enhance one's status in society, given the esteem with which learning was regarded in most Asian countries. 'Minke', the Javanese boy in Toer's novel, has learned Dutch, which brings him to the notice of the Dutch Assistant Resident, the highest official in the district. Suddenly, 'Minke' finds that he has become an object of rapt and respectful attention (source 6.11).

Furthermore, a prolonged exposure to Western education tended to erode its recipients' faith in inherited learning, particularly if it did not conform to the laws of logic or the revelations of science. 'Minke' was affected in this way (source 5.15), and he never left Java!

Imagine the impact on young Asian minds of actually going to the West and seeing its marvels face to face? Most of those who did travel to the West, whether as tourists or students, came back like Bihari Kayastha Mahesh Charan Sinha (source 6.12), with their cultural values dramatically transformed.

Source 6.10 The Colombo Academy, 1848

Those who've read Boyd's Anthon's Horace
And Valpy's Vergil need not fear;
Sir Colin promises to give you
Six times six good pounds a year.

And did you read the books of Euclid?
Know you to count the stars? Ah dear!
Sir Colin says that you may get then
Six times six good pounds a year.

Then join you all, off with your caps, boys,
And let Sir Colin have a cheer;
Hurrah, hurrah, you'v won the world, boys,
Six times six good pounds a year.

S Goonatilake, p. 97.

Question • Is the anonymous author of this poem praising the Colombo Academy or ridiculing it?

Source 6.11 'Minke' discovers the worth of an official invitation

That morning I set off to the post office. The postmaster, I don't know his name, an Indo, shook my hand and praised my Dutch at the previous night's reception as being very excellent and exact. All the office employees stopped working just to listen to our conversation and to take in what I looked liked.

'We would be very proud if you would work here; you are an H.B.S. student, yes?'

'I only want to send a telegram,' I answered.

'There's no bad news, I hope?'

'No.'

The postmaster attended to me himself and gave me the form. He invited me to sit at a table, and I began to write, then I handed the form back to him. Once again he attended to it himself.

'If you have the opportunity, perhaps we could invite you to dinner?'

It appeared that the Assistant Resident's invitation had become big news in B. It could be predicted that all the officials, white and brown, would be sending me invitations. So, all of a sudden, I'd become a prince without a principality. How tremendous, an H.B.S. student! in his last year! in the middle of an illiterate society. They will all be out to indulge me. If the Assistant Resident has started inviting you, naturally you are without flaw, everything you do is right, there is nothing you would ever do that could be said to have violated Javanese custom.

P Toer, p. 121.

Questions

- What was an 'Indo'?
- What was an 'Assistant Resident'?

street scene in ·andigarh, India. The ·arding proudly ·nounces that Professor ·ehta is an 'England-·urned', someone to looked up to.

Source 6.12 Mahesh Charan Sinha on the enlightening experience of foreign travel, 1907

[T]he extent of degradation which India has now reached, how deeply superstition has permeated every inch of Indian ground, cannot be realized unless one goes out of India, and inspects things personally, and compares the people of India with those of other countries. It is then alone possible to find out how narrow-minded and extremely short-sighted are the views and practices of those that pass for educated in India.

These men, utterly unconscious of their ignorance and superstition, attach the highest importance to foolish frivolities. With them what to eat, where to eat, at whose hands to eat, is all important—foolish considerations for which the rest of the globe has not a moment's thought to bestow. Oh! what a pity that the intelligent Hindus who can unravel the most complex mysteries of Metaphysics should be so perverted as not to be able to understand the simplest of propositions that if the breaking of caste and the removal of such absurdities as forbidden land and forbidden food has not injured the whole world, how can it injure Hindus alone who form but a fractional part of the whole world.

L Carroll, pp. 294–5.

Question • How did Sinha's experience of the West compare with that of other Asian travellers you have read about?

Asians and Christianity

One of the major arenas of East–West contact during the age of imperialism (and one of the ongoing themes of this book) was that of religious proselytisation. From the very start, missionaries were in the forefront of the European forces which headed East. They penetrated the interior of countries like China to a far greater extent than any other group of whites. Thus, for many Asians, missionaries and the religion of Christianity they brought with them, came to symbolise the essence of the Western challenge to their traditional culture. How did they respond?

Asian élites generally had the most to lose from the spread of Christianity (with its subversive notion of equality before God), and so were, not surprisingly, fairly hostile. Rulers especially reacted strongly to what they perceived as an initial threat to their **secular** power (source 6.13). At the same time the local priestly class was angered by the missionaries' blasphemous attacks on their gods and rituals. Sir Saiyyid Ahmad Khan, a devout Muslim, points out that this tactic contravened the spirit of tolerance which distinguished Asian religions such as Hinduism and Buddhism (source 6.14).

secular relating to worldly affairs rather than religious ones

But the response of the masses was more divided. Some, urged on by their priests, and stirred up by rumours that the missionaries were involved in terrible practices like the kidnapping and eating of orphaned children, showed their anger in riots, assaults and destruction of missionary property. However, others responded positively to the appeal of the missionaries and converted to the new faith.

Raised in the more tolerant and easy-going atmosphere of the late 20th century, most modern-day Western historians find it difficult to empathise with the missionary cause, and have asserted that the former was more characteristic. Yet the American **evangelist** W A P Martin may have a point when he argues that anti-missionary protests in China were motivated by a dislike of foreigners in general rather than of Christianity in particular (source 6.15).

evangelist lay preacher of the gospel

Source 6.13 Edict of the Vietnamese Emperor Minh Mang, 1833

I, Minh-Mang, the king, speak thus. For many years men from the Occident have been preaching the religion of Dato and deceiving the public, teaching them that there is a mansion of supreme bliss and a dungeon of dreadful misery. They have no respect for the God Phat and no reverence for ancestors. That is great blasphemy indeed. Moreover, they build houses of worship where they receive a large number of people, without discriminating between the sexes, in order to seduce the women and young girls; they also extract the pupils from the eyes of sick people. Can anything more contrary to reason and custom be imagined?

Last year we punished two villages steeped in this depraved doctrine. In so doing we intended to make our will known, so that people would shun this crime and come to their senses.

Now then, this is our decision: although many people have already taken the wrong path through ignorance, it doesn't take much intelligence to perceive what is proper and what is not; they can still be taught and corrected easily. Initially they must be given instruction and warnings, and then, if they remain intractable, punishment and pain.

Thus we order all followers of this religion, from the mandarin to the least of the people, to abandon it sincerely, if they acknowledge and fear our power. We wish the mandarins to check carefully to see if the Christians in their territory are prepared to obey our orders and to force them, in their presence, to trample the cross underfoot. After this they are to pardon them for the time being. As for the houses of worship and the houses of the priests, they must see that these are completely razed and, henceforth, if any of our subjects is known to be guilty of these abominable customs, he will be punished with the last degree of severity, so that this depraved religion may be extirpated.

This is our will. Execute it. 12th day of the eleventh moon, 13th year of our reign.

[the royal seal]

SECRET ANNEX TO THE EDICT

The religion of Jesus deserves all our hatred, but our foolish and stupid people throughout the kingdom embrace it *en masse* and without examination. We must not allow this abuse to spread. Therefore we have deigned to post a paternal edict, to teach them they must correct themselves.

The people who follow this doctrine blindly are nonetheless our people; they cannot be turned away from error in a moment. If the law were followed strictly, it would require countless executions. This measure would cost our people dear, and many who would be willing to mend their ways would be caught up in the proscription of the guilty. Moreover, this matter should be handled with discretion, following the [Confucian] maxim, which states: 'If you want to destroy a bad habit, do so with order and patience,' and continues: 'If you wish to root out an evil breed, take the hatchet and cut the root.'

We order all the *tong doc* and all others who govern:

1. Carefully to attend to the instruction of their inferiors, mandarins, soldiers, or populace, so that they may mend their ways and abandon this religion;
2. To obtain accurate information about the churches and homes of missionaries, and to destroy them without delay.
3. To arrest the missionaries, taking care, in doing so, to use guile rather than violence; if the missionaries are French, they should be sent promptly to the capital, under the pretext of being employed by us to translate letters. If they are indigenous, you are to detain them in the headquarters of the province, so that they may not be in communication with the people and thus maintain them in error. Take care lest your inferiors profit from this opportunity by arresting Christians indiscriminately and imprudently, which would cause trouble everywhere. For this you would be held guilty. . . .

You, provincial prefects, act with caution and prudence, do not stir up trouble; thus you will make yourselves worthy of our favor. We forbid this edict to be published, for fear that its publication might cause trouble. As soon as it reaches you, you alone are to acknowledge it. Obey.

M E Gettleman (Ed), pp. 34–6.

Question • Why was the second part of this document (the annex) not made public?

Source 6.14 Sir Saiyyid Ahmad Khan on the alarming methods of Christian missionaries, 1858

The missionaries moreover introduced a new system of preaching. They took to printing and circulating controversial tracts, in the shape of questions and answers. Men of a different faith were spoken of in a most offensive and irritating way. In Hindustan these things have always been managed very differently. Every man in this country, preaches and explains his views in his own Mosque, or his own house. If any one wishes to listen to him, he can go to the Mosque, or house, and hear what he has to say. But the missionaries' plan was exactly the opposite. They used to attend the places of public resort, markets for instance, and fairs where men of different creeds were collected together, and used to begin preaching them. It was only fear of the authorities that no one bid them off about their business. In some districts the missionaries were actually attended by policemen from the station. And then the missionaries did not confine themselves to explaining the doctrines of their books. In violent and unmeasured language they attacked the followers and the holy places of other creeds: annoying, and insulting beyond expression the feelings of those who listened to them. In this way too the seeds of discontent were sown deep in the hearts of the people.

S Ahmad Khan, *Cause*, p. 126.

Question • How might a missionary have answered Sir Saiyyid's criticisms?

Source 6.15 In defence of China missions, *c.* 1900

But *are* the people unwilling to have missionaries live among them? If they were we should have had to count many more than twenty riots during this quarter of a century. Their increase has not kept pace with the growth of the missionary work. One a year in a country of such vast extent, and with a missionary force of over two thousand, is no proof of popular ill-will, but rather the reverse.

The impression made by these riots is the more profound, as, in addition to sporadic manifestations, they occasionally burst forth with the virulence of an epidemic. The study of these epidemics will show the nature of the disease. In 1891 four such outbreaks occurred. They were all on the banks of the Yang-tse, and all at ports of trade, nor were they, save in one instance, specially aimed at missionaries. Of the hundreds of missionaries living away from the river, scarcely one was molested. It is morally certain that, among the mixed motives of the excited masses, the diversion of the carrying trade from native junks to foreign steamers was at the bottom of the movement. On the Upper Yang-tse, where two of the riots occurred, so strong was the opposition to steamers ascending the rapids that the British minister felt constrained to waive the exercise of that right. No special effort was made to keep missionaries out of Chungking, but the mandarins moved heaven and earth to prevent the coming of the steamer 'Kuling.'

A few years ago a Hindu soldier on guard at the British consulate at Chinkiang struck a Chinaman. In half an hour all the foreign houses in the settlement were laid in ashes. At Canton a foreign tide-waiter in the customs service shot a boy by accident. A furious attack was made on the foreign quarter, which narrowly escaped destruction. At Ichang, in 1895, a shot from an air-gun striking a small official, the populace threw themselves on the handful of foreigners, and a massacre would have ensued but for the opportune arrival of a force from a gunboat. These instances (and such are numerous) suffice to show what fires are burning beneath a thin crust of cold lava, and to prove that if missionaries are attacked oftener than others it is chiefly because they are more exposed.

W A P Martin, pp. 446–7.

Question • Does Martin's argument prove that the Chinese were really well disposed towards the Christian missionaries?

Cultural Cringe

Westernisation brought many beneficial changes to Asian life, but it also caused serious social problems. Certain wealthy élites, for example, became obsessed with Western technology, and spent literally fortunes indulging their appetites. A case in point was the **Francophile** monarch Norodom of Cambodia (*r.*1859–1904). Virginia Thompson notes in her history of Indo-China that he extravagantly squandered a large slice of the budget on mechanical gadgets and tasteless palaces in mock-European style (source 6.16).

Western education also produced mixed results. While many Asians picked up hard-won and valuable skills, others acquired only a taste for liquor and an inflated sense of their own importance. According to the Spanish Jesuit Eduardo Navarro, who visited Manila in 1897, the Philippines was overrun with dropouts from the universities who were vain, arrogant and idle (source 6.17). Likewise, according to the Bengali nationalist Narendranath Sen, English education in India produced a **hybrid** generation possessing all the vices of the British but none of their virtues (source 6.18).

rancophile lover f French people nd things

ybrid thing omposed of ifferent elements; ross-breed

Source 6.16 King Norodom's mania for gadgets

Norodom incarnated Oriental wiles and childishness in his reaction to the new Western influences. He adored every new gadget and he mixed them indiscriminately with Cambodian objects. He wore a series of semi-military uniforms studded with precious stones; his palaces were filled with mechanical toys; he had a statue of Napoleon III decapitated and his own sculptured head substituted—symbolic of the superimposing, without amalgamation, of one civilization upon another. His subjects naturally followed the royal example in so far as they were able. His Ministers took up bicycling with ardour until a new Resident Superior appeared with a motor-car, so they abandoned cycling and bought cars, only substituting gentle gongs for the raucous Klaxon. The king built a brick villa in the worst possible European taste, and they did likewise. An important by-product of this inrush of bad Occidentalism was the decline of the national Cambodian arts. An amusing picture of Norodom in 1881 has been left by Rivière when the king—then in his forty-seventh year and the father of only seventy-two offspring—visited him on board his warship. Norodom was enchanted with the twenty-one-gun salute, but even more by the portable roulette wheel the commandant had brought with him. The king spent such a delightful evening that he accepted without demur all the French demands. Under such casual auspices was the Protectorate launched.

V Thompson, pp. 358–9.

Question • What does the author mean by 'bad Occidentalism'?

Source 6.17 Pretensions of Filipino university students, *c.* 1897

The attention of every studious and observing man, who has lived in residence in the Filipino provinces, is not a little struck by the excessive number of young men, who having taken more or less courses in Manila, but without concluding the course begun, or even taking the degree of bachelor, after their parents have spent considerable sums on them, return to their villages with very little or no virtue, but with many vices. At first sight one notes in these young men an irritating radical attitude and a freedom mixed with unendurable arrogance and vanity. Their fellow countrymen, whom they disdain because they possess, although in a superficial manner, the Castilian speech full of phrases and sounds, which would make the most reserved Viscayan laugh, and of high-sounding words which they use without understanding their real significance, immediately look up to them as so many Senecas [great orators]. They are persuaded that they are perfect gentlemen, for by dint of seeing them practiced they have learned a few social formulas; they wear a cravat, and boots, and pantaloons of the latest style. For the rest, they are completely devoid of fundamental knowledge, and of the fundamentals of knowledge in the studies which they have taken, and have acquired only a slight tint of the part, let us say the bark of those studies, which they conclude by forgetting in proportion as time passes and their passions increase. These young men who forget what they have learned with so great facility, do not, as a general rule, devote themselves to any work, for they do not like work and cannot perform any; for the habits that they have contracted are very different—habits of pastime, idleness, and the waste of their paternal capital.

E H Blair and J A Robertson (Eds), pp. 315–6.

Question • Is this observer suggesting that Western education in the Philippines has failed?

Source 6.18 Denationalisation in India, 1883

Through the action of a purely English education, we have lost faith in our religion without getting something better in substitution; we have contracted more the vices than the virtues of the Englishman; we have got merely an external polish, while we are rotten within; we have developed more our physical than our spiritual nature; and many other evils have been brought in, which would not have come into existence at all, if Western education had gone hand in hand with Eastern education. To be brief, our nationality and our spirituality, the two most important elements which contributed so much to the glory of Ancient India, have departed.

***Indian Mirror*, 20 May 1883.**

Question • What English 'vices' could Indians have 'contracted' through Wester education?

Loneliness of the Alienated Intellectual

Anglophile lover of English people and things

Perhaps the most damaging aspect of Westernisation, however, was the way it dislocated Asians from their roots and stranded them in a sort of cultural limbo between West and East. Indonesian nationalist Sutan Sjahrir (1909–66) was educated abroad in Leiden. When he came home, he felt much closer to Europeans than to his own people, as he frankly confesses in a letter written from gaol in 1934 (source 6.19a). India's Jawaharlal Nehru (1889–1964), who received a very **Anglophile** education at Harrow School and Cambridge University, likewise admits in his autobiography to a sense of being marooned, 'out of place everywhere, at home nowhere' (source 6.19b).

Jawaharlal Nehru (1889–1964) (right) with his father Motilal, sister 'Nan', wife Kamala, and daughter Indira. What office did Indira fill in later life?

Thus, the great challenge which confronted Asian thinkers during the later colonial period was to find a way of marrying modernity with tradition, of reconciling their intellectual commitment to the West with their sentimental attachment to family and caste and religion. One solution was to pretend that so-called Western ideas were really Asian ideas which had got twisted in translation. For example, the Chinese Tai'pings preached a garbled version of Protestantism. They claimed that their leader Hung Hsiu-ch'uan (1814–64) was the younger brother of Jesus Christ, ordered by God to return China to the true path from which it had strayed centuries before (source 6.20). Similarly, the Bengali Brahmo Samajist Keshub Chandra Sen, leader of a sect which had been influenced strongly by both Christian ethics and Islamic **monotheism**, attempted to Asianise Christianity by arguing that it was 'founded and developed by Asiatics' (source 6.21).

monotheism doctrine that there is only one God

communism ideal social state in which there is no private property

Another solution was to identify only with those Western traditions which were sympathetic to Asia and hostile to imperialism—such as Marxism, which became influential after the Russian Revolution of 1917. One of the most famous of Asian Marxists was the Vietnamese leader Nguyen Sinh Cung (1890–1969), better known as Ho Chi Minh. In source 6.22 he recalls how, while living in France during the First World War, he read Lenin. At this time he became convinced that only **communism** could 'liberate the oppressed nations . . . from slavery'.

Source 6.19 Nehru and Sjahrir: 'Estranged' nationalists

a Am I perhaps estranged from my people? Why am I vexed by the things that fill their lives, and to which they are so attached? Why are the things that contain beauty for them and arouse their gentler emotions only senseless and displeasing for me? . . . We intellectuals here are much closer to Europe and America than we are to the Boroboedoer or Mahabharata or to the primitive Islamic culture of Java and

Sumatra. Which is our basis: the West, or the rudiments of feudal culture that are still found in our Eastern society?

S Sjahrir, pp. 66–7.

b Indeed, I often wonder if I represent any one at all, and I am inclined to think that I do not, though many have kindly and friendly feelings toward me. I have become a queer mixture of the East and the West, out of place everywhere, at home nowhere. Perhaps my thought and approach to life are more akin to what is called Western than Eastern, but India clings to me, as she does to all her children, in innumerable ways; and behind me lie, somewhere in the subconscious, racial memories of a hundred, or whatever the number may be, generations of Brahmans. I cannot get rid of either that past inheritance or my recent acquisitions. They are both part of me, and, though they help me in both the East and the West, they also create in me a feeling of spiritual loneliness not only in public activities but in life itself. I am a stranger and alien in the West. I cannot be of it. But in my own country also, sometimes, I have an exile's feelings.

J Nehru, p. 596.

Question • When these musings were written Nehru and Sjahrir had not met and probably did not know of each other's existence, yet the two extracts are very similar. What does this tell us about the historical processes at work in Asia in the early 20th century?

Source 6.20 Tai'ping Christianity, *c.* 1852

Again it has been falsely said that to worship the Great God is to follow barbarians' ways. They do not know that in the ancient world monarchs and subjects alike all worshiped the Lord God. As for the great Way of worshiping the Lord God, from the very beginning, when the Lord God created in six days Heaven and earth, mountains and seas, man and things, both China and the barbarian nations walked together in the great Way; however, the various barbarian countries of the West have continued to the end in the great Way. China also walked in the great Way, but within the most recent one or two thousand years, China has erroneously followed the devil's path, thus being captured by the demon of hell. Now, therefore, the Lord God, out of compassion for mankind, has extended his capable hand to save the people of the world, deliver them from the devil's grasp, and lead them out to walk again in the original great Way.

W T de Bary (Ed), *Chinese Tradition*, p. 688.

Question • How did Tai'ping Christianity differ from orthodox Protestantism?

Source 6.21 Keshub Chandra Sen on the 'Asiatic' Christ, 1866

And was not Jesus Christ an Asiatic? [Deafening applause.] Yes, and his disciples were Asiatics, and all the agencies primarily employed for the propagation of the Gospel were Asiatic. In fact, Christianity was founded and developed by Asiatics, and in Asia. When I reflect on this, my love for Jesus becomes a hundredfold intensified; I feel him nearer my heart, and deeper in my national sympathies. Why should I then feel ashamed to acknowledge that nationality which he acknowledged? Shall I not rather say he is more congenial and akin to my Oriental nature, more agreeable to my Oriental habits of thought and feeling? And is it not true that an Asiatic can read the imageries and allegories of the Gospel, and its descriptions of natural sceneries, of customs, and manners, with greater interest, and a fuller perception of their force and beauty, than Europeans. [Cheers.] In Christ we see not only the exaltedness of humanity, but also the grandeur of which Asiatic nature is susceptible. To us Asiatics, therefore, Christ is doubly interesting, and his religion is entitled to our peculiar regard as an altogether Oriental affair. The more this great fact is pondered, the less I hope will be the antipathy and hatred of European Christians against Oriental nationalities, and the greater the interest of the Asiatics in the teachings of Christ. And thus in Christ, Europe and Asia, the East and the West, may learn to find harmony and unity. [Deafening applause.]

W T de Bary (Ed), *Indian Tradition*, p. 620.

Question • What do you think of Keshub's assertion that Jesus Christ was an 'Asiatic'?

Source 6.22 Ho Chi Minh recalls his conversion to Leninism

Heated discussions were then taking place in the branches of the Socialist Party, about the question whether the Socialist Party should remain in the Second International, should a Second-and-a-half International be founded or should the Socialist Party join Lenin's Third International? I attended the meetings regularly, twice or three times a week and attentively listened to the discussion. First, I could not understand thoroughly. Why were the discussions so heated? Either with the Second, Second-and-a-half or Third International, the revolution could be waged. What was the use of arguing then? As for the First International, what had become of it?

What I wanted most to know—and this precisely was not debated in the meetings—was: which International sides with the peoples of colonial countries?

I raised this questions—the most important in my opinion—in a meeting. Some comrades answered: It is the Third, not the Second International. And a comrade gave me Lenin's 'Thesis on the national and colonial questions' published by *l'Humanité* to read.

There were political terms difficult to understand in this thesis. But by dint of reading it again and again, finally I could grasp the main part of it. What emotion, enthusiasm, clear-sightedness, and confidence it instilled in me! I was overjoyed to tears. Though sitting alone in my room, I shouted aloud as if addressing large crowds: 'Dear martyrs, compatriots! This is what we need, this is the path to our liberation!'

At first, patriotism, not yet Communism, led me to have confidence in Lenin, in the Third International. Step by step, along the struggle, by studying Marxism–Leninism parallel with participation in practical activities, I gradually came upon the fact that only Socialism and Communism can liberate the oppressed nations and the working people throughout the world from slavery.

M E Gettleman (Ed), pp. 37–8.

Question • What was the basis of the conflict between the 'Second International' and the 'Third International'?

7 Growth of Asian Nationalism

In imperialism, nothing fails like success.

William Ralph Inge, 19

THIS CHAPTER SHOWS HOW the stimulus of Western ideas led naturally and inevitably to the growth of national consciousness among educated Asian élites. At the same time the increasing discrepancy between what colonialism promised and what it actually delivered caused more and more of these idealistic young men and women to become dissatisfied and alienated from their European masters.

KEY DATES

1859 Peasants riot in Bengal against oppressive conditions associated with the forced cultivation of indigo.
1874 Surendranath Banerjea (1848–1926) dismissed from the Indian Civil Service for a trifling misdemeanour.
1885 The Indian National Congress meets for the first time in Bombay.
1893 Swami Vivekananda (1865–1902) wins international acclaim at the first World Parliament of Religions in Chicago.
1896 Jose Rizal (b. 1861), respected Filipino eye-surgeon and novelist, executed by the Spanish authorities for sedition.
1907 *Nacionalista* Party formed in the Philippines.
1908 Poona journalist Bal Gangadhar Tilak (1856–1920) sentenced to eight years jail for incitement to murder.
1911 Raden Adjeng Kartini (1879–1904) publishes her autobiographical *Letters of a Javanese Princess*.
1912 Unsuccessful attempt to assassinate the Viceroy of India Lord Hardinge (1858–1944) during a procession in Delhi.
1919 Chinese students in Peking demonstrate against the Versailles peace settlement.
1920 M K Gandhi (1869–1948) launches a movement of 'non-co-operation' designed to topple the British Raj in India.
1927 Dutch-trained engineer Sukarno (1901–1970) founds the Partai Nasional Indonesia (Indonesian Nationalist Party).

ISSUES AND PROBLEMS

Nationalism is the creed which holds that the nation-state is the highest and most perfect form of political organisation so far devised by man. Believing that loyalty to the nation-state supersedes all other loyalties, nationalism justifies and promotes the activity of nation-building. But how are nations 'built'? Do they come ready-packaged, like kit houses, with a set of instructions for assembly, or must they be constructed from the ground up, brick by brick? Looking around the modern world, it is apparent that the people comprising a nation share certain common characteristics. For instance, they usually speak the same language, have the same broad racial characteristics, and profess the same religion. However, it is equally obvious that none of today's 'nations' wholly fit this prescription. Australia, the United States and many other 'new world' countries are inhabited by peoples of diverse ethnic background. 'Hindu' India has 40 million Muslims and 10 million Sikhs. Switzerland boasts three 'national' languages. A single culture, therefore, is not an essential prerequisite for nationhood. Nations can be built out of all sorts of odd raw materials. Fundamentally, all that is needed is a territory (the 'homeland'), an agreed goal, a common pool of political values (for example democracy), and lots of determination.

Nations, then, are made, not born. But why? What is so much better about a nation than a village, a territorial clan, or a dynastic empire? Clearly, what we would now term ethnic groups have existed in Asia for centuries. Yet these groups did not recognise themselves as such. Rather, they formed part of other (smaller and larger) social and political organisations. What catalyst caused Asians suddenly to see their world differently, and prompted them to seek to redraw the map of the Asian continent along European lines?

prima facie based on first impression

Prima facie, the trigger was the European conquest which was accompanied by new political ideas. However, that does not explain why Asians took to nationalism so enthusiastically. As the previous chapter showed, Western notions were not always accepted uncritically. Furthermore, most national movements in Asia were strongly anti-colonial in character, because the idea of the nation-state carried overtones of popular sovereignty and representative government. How did Asian nationalists reconcile their bitter opposition to European rule with their fervent commitment to what was, in essence, a European institution?

Another indication that nationalism in Asia was not simply an instinctive reaction to imperialism is the fact that it arose relatively late in the piece. Europeans had been in Asia since 1500; the Spanish had ruled the Philippines since 1580; and by the late 18th century the British and the Dutch were in firm control of large areas of India and Indonesia.

Yet it was not until the end of the 19th century that the first stirrings of nationalism began to be heard: the Indian National Congress was formed in December 1885; the Nacionalista Party of the Philippines in 1907; Budi Utomo, the first modern Indonesian political organisation, in 1908; the Chinese Kuomintang in 1912; and the Ceylon National Congress in 1919.

Why did it take 100 years and more for the doctrine of nationalism to be established? Why, precisely at this time, did the networks of collaboration so carefully nurtured by the colonial regimes begin to break down? Was it because of the increasing pace of European economic penetration, or changing administrative priorities (for example, the Dutch Government's swing, around 1900, to an Ethical Policy in Indonesia)? Or did external events such as the 1917 Bolshevik revolution in Russia or the struggle for 'home rule' in Ireland cause the breakdown?

THE DEBATE

There are broadly three schools of thought about the rise of Asian nationalism. The first, composed largely but not exclusively of European writers, believes that Asian nationalism developed in response to the modernising impact of colonial rule. According to this school, colonialism laid the basis for the development of national consciousness by knitting together previously separate regions, actively promoting Western languages and studies (which gave Asians access to European political ideas) and creating new opportunities for social mobility.

The second school, while agreeing that colonialism acted as a catalyst, prefers to stress its negative side. In their view, nationalism arose as a natural reaction to the oppression, exploitation, and racial discrimination which flourished under European rule. As Helen Lamb says of French Indo-China:

> It was a vicious circle: the greater the resistance, the more high-handed and brutal the repression, and the more costly to France in lives and money. This in turn led to grinding more taxes out of the Vietnamese people so as to make the colony a paying proposition. But these same fiscal measures precipitated further dissatisfaction among the Vietnamese. As a result, this early and unsuccessful resistance, since it etched more and more deeply the ugly face of colonialism, helped to set in motion the very forces that led to renewals of the struggle and thus to ultimate victory over the French.[1]

In the writings of the third school, however, the European dimension is played down. Instead, Asian nationalism is portrayed as emerging as a by-product of a renewed pride in traditional cultures. Briefly, their argument runs as follows. Since the instinct for self-government, for freedom, is basic to all peoples, Asians didn't need to be taught it by the West. On the other hand, concepts like 'democracy' and 'individual liberty' were foreign, and made little sense to villagers whose lives were a constant battle with hunger. Consequently, nationalism in the East made headway, as a mass movement, only when it was translated (by leaders such as 'Mahatma' Gandhi) into traditional (usually religious) idioms.

Effects of Western Education

Asian nationalist leaders were generally drawn, as American journalist John Gunther remarked in 1939, from the Western-educated sections of the population (source 7.1). This was not coincidental.

On the whole, this group was fairly affluent and high caste (in India, the large majority of 19th century Congress leaders were Brahmins). Furthermore, they had been more exposed than any other group to the heady doctrines which the 18th century 'Enlightenment' and the American and French Revolutions had given birth to in Europe: ideas of equality and liberty, of free speech and **popular sovereignty**. Take, for example, Bengali Nirad Chaudhuri (b. 1897). Though from a traditional high-caste household, Chaudhuri's boyhood heroes were all European. It was only after 1905, when he was in high school, that the Young Turk Revolution—itself Western-inspired—provided him with a role model closer to home (source 7.3).

popular sovereignty government by people's elected representatives

But what of the masses, most of whom did not even know how to read? How did they come to acquire the sense of being part of a single nation? One answer is suggested by missionary James Long's testimony before the Commission set up to enquire into the so-called 'Blue Mutiny' of 1859 in Bengal. The mutiny involved a series of riots among Bengali peasants protesting at the conditions under which they were compelled to grow indigo. According to Long, newspapers in **vernacular** languages were beginning to penetrate even remote corners of the countryside; read aloud, to those who could not read for themselves, these newspapers were bringing the message of nationalism to an ever-widening circle of the common people (source 7.2).

vernacular language or dialect of a particular group

Source 7.1 John Gunther's impression of Asian nationalists, 1939

It is a striking phenomenon, as Frances Gunther has pointed out, that native peoples tend to take on the complexion of the folk that rule them. Human and political nature *does* change—at least in the East. For instance, the Filipinos, after forty years of Americanisation, are more American than the Americans. A barman in the Hotel Manila, a deputy in the Filipino legislature, could have lived all their lives on Broadway so far as general attitude is concerned. Similarly in British India most of the great nationalist leaders behave strikingly like Englishmen. They go to school in England; they learn about that extraordinary concept, 'fair play', and they look at 'gentlemen'; back in India, they become English gentlemen themselves, even when bitterly attacking British rule. The same thing is true in the Netherlands Indies. Most of the nationalist leaders I met were almost indistinguishable from Dutchmen.

J Gunther, p. 368.

Question • To the extent that Gunther was right, did this superficial resemblance between the nationalists and their colonial masters help, or hinder, the cause of nationalism?

Source 7.2 Evidence of Reverend James Long before the Indigo Riots Commission, 1880

. . . my own enquiries and duties have brought two causes prominently to my notice, as conducing to independence of mind among the masses; first, English education, happily spreading in the country among the natives, is giving them a sense of freedom, leavening their minds with a regard to a sense of justice, and imparting to them an English tone of revulsion against oppression. It is also welding the natives of the different Presidencies into one patriotic mass, with a community of feeling on Indian subjects. Thus a native of Calcutta, on a recent visit to Bombay, was enabled to address numbers of Parsees and Guzeraties in English; though they knew nothing of each other's vernacular. A pamphlet was published by a native in this city, some time ago, in English, and was reprinted by his countrymen in Madras and circulated widely. Madras and Bombay, like Calcutta, have newspapers in English, conducted by natives, and advocating the views of educated natives.

This influence is radiating downwards. The substance of those newspapers and pamphlets in English are being communicated orally, or by means of translations to the masses of the people.

The vernacular press is rising into great importance, as a genuine exponent of native opinion, and it is to be regretted that the European community pay so little regard to its admonitions and warnings. It is the index of the native mind. In 1853, I visited Delhi, Agra, and Lucknow, and particularly examined the statistics connected with the vernacular press, in the Upper Provinces, and I remember the impression with which I left Delhi, after I had been through its lanes and gullies, exploring the localities of its vernacular presses. I felt then very strongly, how little the Europeans of Delhi and other cities were aware of the prodigious activity of the vernacular periodical press, and the impression it was evidently producing on the native mind as tested by the avidity with which books, treating on native and political subjects, were purchased.

J R McLane (Ed), pp. 32–6.

Question • What does Long (a) like and (b) dislike about British colonialism in Bengal?

Source 7.3 From *The Autobiography of an Unknown Indian*

In the actual unfolding of contemporary history it made us read with delight and high hopes the news of the political revolutions of our youthful days—the Russian Revolution of 1905, the Young Turk Revolution of 1909, and the Chinese Revolution of 1911. We invariably identified political freedom with two things: the absence of an absolute monarch and the presence of an assembly of representatives of the people. But we never worked out the relationship between these representatives and the general mass of the population of a country.

Certain modern personalities and movements contributed powerfully to our political consciousness, of which there were two clearly discernible facets. The first and rational facet was indoctrinated by Burke and Mill, but shaped in its practical expression by the liberalism of Gladstone and Lincoln. The second facet was purely emotional, and its inspiration was furnished by Rousseau and Mazzini besides the Ancients. The methods of political action were suggested by the leaders of the American Revolution, the Italian Risorgimento—particularly Garibaldi—and the Irish Nationalists. The entire course of English constitutional history and, more especially, the turmoils of the seventeenth century, together with the American, French, Italian, and Irish movements were freely drawn upon for precedents and also for operational hints.

N C Chaudhuri, p. 227.

Question • What does the writer mean by 'precedents' and 'operational hints'?

Oppression Frustrating Asians' Expectations

At the same time as the growth of a national consciousness, there occurred a dramatic change in Asian attitudes towards colonialism. Initially, as chapter 5 showed, Asian élites, especially, welcomed the coming of European rule, which ushered in an era of relative peace and prosperity via modernisation.

However, as the 19th century wore on, this good relationship with the West began to turn sour. The peoples under colonial rule gradually began to turn against their self-styled benefactors. By the end of the century, nationalist leaders in Ceylon, India, and the Philippines were starting to demand their freedom. Why did this change occur?

For those Asians who had been exposed to Western education, disenchantment with colonialism often grew out of personal setbacks or slights at the hands of Europeans. Bengali Surendranath Banerjea sailed for England in 1869 to prepare for the competitive examination which led to entry to the Indian Civil Service. At the time he was thrilled by the prospect that, in a few years, he would become one of the first of his race to qualify for that privileged ***corps d'élite***. When the exam results went up, he was not disappointed, and he gained a posting as Assistant Collector in a district in the north of his home province. But then things began to go wrong. Two years into his new job, Banerjea was found guilty of a minor offence and dismissed from the service. (Had a white man committed the offence, at worst he would have received a severe reprimand.) Angry but still resolved to succeed in the white man's world, Banerjea returned to London and studied to be a barrister. Again he passed the exams with flying colours, only to be informed that he would not be called to the Bar (source 7.4). Is it any wonder that this young man felt betrayed by the system? And Banerjea was not alone.

corps d'élite
select group

All over Asia, petty acts of racial oppression prevented bright, Westernised Asians from realising their potential and satisfying their ambition. Source 7.5, from Raden Adjeng Kartini's *Letters of a Javanese Princess*, describes an example of such oppressive behaviour.

Source 7.4 The sad case of Surendranath Banerjea, *c.* 1875

I continued eating my dinners, and the time came when I was to be called. That was some time in April or May, 1875. My name was duly put up. An objection was, however, raised, from what quarter or by whom I knew not, nor did I care to enquire then nor do I even now. My dismissal from the Civil Service was considered to be a fatal objection, and the Benchers of the Middle Temple declined to call me to the Bar. An old English barrister, Mr. Cochrane, who for many years was an eminent leader of the Calcutta Bar, warmly interested himself in my case. Old as he was and almost tottering with the weight of years, he did all that was humanly possible. It was a pleasure to see the old man, fired with the enthusiasm of youth on my behalf. He was a grand specimen of a type which I fear is rapidly passing away. But all his efforts were made in vain. From the Civil Service I had been dismissed. From the Bar I was shut out. Thus were closed to me all avenues to the realisation of an honourable ambition.

The outlook was truly dark. My friends declared that I was a ruined man, and that there was no hope for me on this side of the grave. Even the great Kristo Das Pal, editor of the *Hindoo Patriot*, took the same view. A friend, now dead, who achieved considerable distinction as a member of the Calcutta Bar, advised me in a sympathetic vein that I should change my name, go to Australia and seek out a career there for myself. I listened to these friendly counsels with all the equanimity I could muster, but I never despaired, nor even was the exuberant joyousness of my youthful nature darkened by the heavy clouds that lay thick around me. In the iron grip of ruin I had already formed some forecast of the work that was awaiting me in life. I felt that I had suffered because I was an Indian, a member of a community that lay disorganized, had no public opinion, and no voice in the counsels of their Government. I felt with all the passionate warmth of youth that we were helots [serfs], hewers of wood and drawers of water in the land of our birth. The personal wrong done to me was an illustration of the helpless impotency of our people. Were others to suffer in the future as I had suffered in the past? They *must*, I thought to myself; unless we were capable as a community of redressing our wrongs and protecting our rights, personal and collective. In the midst of impending ruin and dark, frowning misfortune, I formed the determination of addressing myself to the task of helping our helpless people in this direction.

S Banerjea, pp. 30–1

Question • How does Banerjea rationalise his personal misfortunes?

Source 7.5 The miseries of a gifted Javanese student, *c.* 1890

The Hollanders laugh and make fun of our stupidity, but if we strive for enlightenment, then they assume a defiant attitude towards us. What have I not suffered as a child at school through the ill will of the teachers and of many of my fellow pupils? Not all of the teachers and pupils hated us. Many loved us quite as much as the other children. But it was hard for the teachers to give a native the highest mark, never mind how well it may have been deserved.

I shall relate to you the history of a gifted and educated Javanese. The boy had passed his examination, and was number one in one of the three principal high schools of Java. Both at Semarang, where he went to school, and at Batavia, where he took his examinations, the doors of the best houses were open to the amiable schoolboy, with his agreeable and cultivated manners and great modesty.

Everyone spoke Dutch to him, and he could express himself in that language with distinction. Fresh from this environment, he went back to the house of his parents. He thought it would be proper to pay his respects to the authorities of the place and he found himself in the presence of the Resident, who had heard of him, and here it was that my friend make a mistake. He dared to address the great man in Dutch.

The following morning notice of an appointment as clerk to a controleur in the mountains was sent to him. There the young man must remain to think over his 'misdeeds' and forget all that he learned at the schools. After some years a new controleur or possibly assistant controleur came; then the measure of his misfortunes was made to overflow. The new chief was a former schoolfellow, one who had never shone through his abilities. The young man, who had led his classes in everything, must now creep upon the ground before the one-time dunce, and speak always high Javanese to him, while he himself was answered in bad

Malay. Can you understand the misery of a proud and independent spirit so humbled? And how much strength of character it must have taken to endure that petty and annoying oppression?

C L M Penders (Ed.), pp. 220–1.

Question • Why do you think the Resident objected to being addressed in Dutch?

Foreign Exploitation of the Natives

Tent-dwellers in Bikaner, India.

Popular resentment against colonialism stemmed in part, as we have seen, from personal disappointment. To this extent its rapid rise in the later 19th century had a lot to do with the fact that Western education was producing more graduates than could be easily absorbed into the professions. This left many young men either with menial jobs or unemployed. But it also reflected changes taking place within the structure of colonialism itself.

As colonial administration expanded its range of activities, it became more intrusive. Regulations increasingly controlled what could or could not be done, and overbearing officials made people's lives a misery by constantly meddling in their affairs. Moreover, bigger government meant higher taxes, which left the poor with even less money of their own. As the anonymous poem cited in source 7.6 eloquently testifies, conditions were deteriorating for ordinary people in Vietnam.

Although it was bad, it was not actually the situation of the Asians that created the unrest. It was the stark contrast between the low standard of living of the masses and the ostentatious life-style of the resident Europeans. By the end of the 19th century a very considerable slice of the Indo-Chinese budget was supporting the lavish salaries paid to a small handful of senior bureaucrats.

At the same time, Europeans working in the commercial sector received significantly higher wages than Asians doing similar jobs (source 7.8). Asian subjects had only to look around them to see that they were being exploited, and they did not need to be told who the culprit was. The radical French playwright and novelist Jean Ajalbert visited Indo-China (ironically, on a grant from the French Government) in the early 1900s. Even among the **proletariat**, he found deep-seated feeling of anger at the damage which colonialism had wrought (source 7.7).

proletariat lowest class of community; wage-earners, especially those who are dependent on daily labour for subsistence

However, it was another 20 years before this sense of outrage spawned a political movement capable of challenging French rule. And another 20 years passed before the nationalists led by Ho Chi Minh were strong enough to proclaim their country's independence (source 7.9).

Source 7.6 French oppression in Vietnam: An anonymous poem, 1900

In the year I Wei, the seventh year of the reign of Emperor Thanh thai [1895],
The war indemnities had to be paid in full and at once.
The official order struck like lightning across the sky.
They came with summons to this village, with rifles to that hamlet.
Every place had to declare the number of its inhabitants, houses, male adults, rice fields.
Taxes were increased greatly and were to be paid in money, not in kind.
With each passing year these taxes mounted,
The cost of all articles rose rapidly, even those of betel, tea, and areca nuts.
The constables, the commissars, the police, the agents of the Security Services, all officials competed to harm the people.
All over the country, city-dwellers
Paid taxes on their persons and their houses.
They had to purchase licenses for peddling.
There were taxes on theaters, singers,
Dogs, pigs, and shops selling mutton.
In their exploitation the French did not miss a single item.
There was a monopoly on salt, and alcohol was stored plentifully in the excise offices.
People were obliged to buy and sell
Or they were accused of being smugglers; the situation was simply wretched.
The laws were iron, in a hundred ways,
Every individual was sorrowful, and every family utterly grieved.
Some people sold their wives, some sold their children.
To sell husbands was no longer remarkable.
How is one to recount the sorrow and suffering?
When one questions Heaven, Heaven remains quite silent.
What debts had our compatriots contracted in their previous lives?
Not only were they exploited by the French, they also suffered a drought.
Of ten crops, more than nine perished.
Then too, there were storms, floods, violent winds, and irregular rains.
Who could tend to his wasted body?
How many died hungry on the sidewalks?
Night and day they were compelled to work for the administration.
No sooner was the younger brother back home than the older brother was at his post,
The administration had a hundred ways of extorting the people's money.
They took collections, they fined and ceaselessly claimed indemnities.
They never checked the truth of their information.
Whenever they heard rumors of unrest anywhere in the country, they immediately sent in their troops.
They governed tyrannically with their laws.

Of superior strength, they oppressed the people.
No one dared complain.
All suffered damage without ever lodging a grievance.
The French gave orders, the Vietnamese obeyed.

B L Truong, pp. 146–7.

Question • Is the value of this document reduced by its anonymity?

A Chinese wall-poster of the 1930s shows China being 'strangled' by a combination of Western capitalism and Japanese militarism.

Source 7.7 The lament of a Vietnamese boatman as heard on the Perfumed River, Hue, *c.* 1910

We have houses of bamboo and straw which the least wind blows down; the mandarins live in pagodas of precious woods covered with tiles, which will withstand the most violent typhoons; the French, oh, the French—they live in splendid palaces where it is warm in the winter and cool in the summer.—We have only one garment of rough cloth, sullied by a season's work and sweat; the mandarins dress in silks in the summer and in velvet during the winter; the French, oh, the French—they have garments of all kinds, made of materials unknown to us.—Our food consists of rough rice seasoned with legumes and fish; the mandarins eat the rarest dishes, fowl, game, the whitest rice; the French, oh, the French—they are not satisfied with what they find in Annam, they import their food from their own country.—Our women, who work all day long, have calloused hands, deformed bodies, and dirty garments; the mandarins have the prettiest girls in the land who are always well groomed and covered with rich gowns; the French, oh, the French—with their money they have all the girls they want, the daughters of mandarins or the daughters of the common people, all let themselves be seduced by them.—We walk only on foot, and often we carry the mandarins in their palanquins; the mandarins do not content themselves with the palanquin of their forefathers; they have rickshas, boats, and even horse carriages; the French, oh, the French—they have means of transportation unknown to us, they travel fast in boats driven by fire, and in carriages big as houses.—We pay taxes to the mandarins and to the French; the mandarins get our taxes and our gifts; the French, oh, the French—they get our taxes, our gifts, and those of the mandarins as well, and take from our land whatever they want.

J Buttinger, pp. 460–1.

Question • Why might one be suspicious of the accuracy of this document?

Source 7.8 Wages of Indian and European employees on the Kolar Goldfields, India, 1900–40

Year	*(a) Average European wage per month (in pounds)*	*(b) Average Indian wage per month (in pounds)*	*Ratio of a/b*	*Real wage index for Europeans (1900 = 100)*	*Real wage index for Indians (1900 = 100)*
1900	17.68	1.44	12.2 : 1	100.0	100.0
1910	24.19	1.52	16.0 : 1	114.3	95.0
1920	37.70	1.89	13.1 : 1	83.6	78.7
1930	44.88	2.42	18.5 : 1	173.0	114.3
1940	46.39	2.72	17.1 : 1	172.7	134.7

C Simmons, 'Labour and Capital', pp. 17–18.

Question • What is a 'real wage'?

Source 7.9 Declaration of independence of the Democratic Republic of Vietnam, 1945

'All men are created equal. They are endowed by their Creator with certain inalienable rights, among these are Life, Liberty, and the pursuit of Happiness.'

This immortal statement was made in the Declaration of Independence of the United States of America in 1776. In a broader sense, this means: All the peoples on the earth are equal from birth, all the peoples have a right to live, to be happy and free.

The Declaration of the French Revolution made in 1791 on the Rights of Man and the Citizen also states: 'All men are born free and with equal rights, and must always remain free and have equal rights.'

Those are undeniable truths.

Nevertheless, for more than eight years, the French imperialists, abusing the standard of

Liberty, Equality, and Fraternity, have violated our Fatherland and oppressed our fellow-citizens. They have acted contrary to the ideals of humanity and justice.

In the field of politics, they have deprived our people of every democratic liberty.

They have enforced inhuman laws; they have set up three distinct political regimes in the North, the Center and the South of Vietnam in order to wreck our national unity and prevent our people from being united.

They have built more prisons than schools. They have mercilessly slain our patriots; they have drowned our uprisings in rivers of blood.

They have fettered public opinion; they have practised obscurantism [concealment of truth] against our people.

To weaken our race they have forced us to use opium and alcohol.

In the field of economics, they have fleeced us to the backbone, impoverished our people, and devastated our land.

M E Gettleman (Ed), pp. 63–4.

Question • What is ironic about the Declaration's reference to the French Revolution?

Colonial Repression of the Natives

As nationalist resentment against the injustices of colonial rule spread and intensified, the colonial authorities tried to control the situation by being harsher, and acts of brutality against protesters and even innocent bystanders multiplied. In India, for example, censorship was imposed, and people were imprisoned without trial. Upon his return from a fact-finding mission in 1907, the Labour Member of Parliament Keir Hardie (1856–1915) was quick to remind the British Government that such acts were totally contrary to what the Westminster system was supposed to represent (source 7.10).

Even when the police were not arresting people, or, as in Amritsar in April 1919, shooting indiscriminately on crowds of peaceful demonstrators (see source 3.16), they were conspicuous by their presence. Simply by being there, they reminded Asians who needed to be reminded that colonialism was essentially an authoritarian regime.

Yet while the police were often brutal and unnecessarily provocative, their conduct was generally less atrocious than that of European businessmen and planters, who were beholden to nobody but themselves. George Orwell spent his early adult years as a rather disenchanted member of the Burma Police Force. In his novel *Burmese Days*, he recreated in fiction the kind of incident that was, as far as he was concerned, far too common: the bashing of a Burmese boy by the European manager of a local company (source 7.11). Source 7.12 is an excerpt from the semi-official Delamarre Report of 1928 which gives a factual description of the way coolies were routinely mistreated on French-owned plantations in Vietnam.

Source 7.10 Keir Hardie's impressions of India, 1908

Recent acts of the authorities, especially the deportation without trial of men of good social standing and position, against whom no charge could be formulated; the suppression of public meetings, the support given to corrupt and inefficient police officials, and the establishment of a secret police service, with its agents in every corner; the growing oppression of the ryots [peasants], and the supercilious way in which their claims for redress are met, are all tending to shake the belief which was formerly universal in the impartiality of British justice and the fairness of British administration.

G Bennet (Ed), p. 358.

Question • Was Hardie an impartial witness?

Source 7.11 White racism in 'Kyauktada', Burma

Ellis wriggled his shoulders—his prickly heat was almost beyond bearing. The rage was stewing in his body like a bitter juice. He had brooded all night over what had happened. They had killed a white man, killed *a white man*, the bloody sods, the sneaking, cowardly hounds! Oh, the swine, the swine, how they ought to be made to suffer for it! Why did we make these cursed kid-glove laws? Why did we take everything lying down? Just suppose this had happened in a German colony, before the War! The good old Germans! They knew how to treat the niggers. Reprisals! Rhinoceros hide whips! Raid their villages, kill their cattle, burn their crops, decimate them, blow them from the guns.

Ellis gazed into the horrible cascades of light that poured through the gaps in the trees. His greenish eyes were large and mournful. A mild, middle-aged Burman came by, balancing a huge bamboo, which he shifted from one shoulder to the other with a grunt as he passed Ellis. Ellis's grip tightened on his stick. If that swine, now, would only attack you! Or even insult you—anything, so that you had the right to smash him! If only these gutless curs would ever show fight in any conceivable way! Instead of just sneaking past you, keeping within the law so that you never had a chance to get back on them. Ah, for a real rebellion—martial law proclaimed and no quarter given! Lovely, sanguinary images moved through his mind. Shrieking mounds of natives, soldiers slaughtering them. Shoot them, ride them down, horses' hooves trample their guts out, whips cut their faces in slices!

Five High School boys came down the road abreast. Ellis saw them coming, a row of yellow, malicious faces—epicene faces, horribly smooth and young, grinning at him with deliberate insolence. It was in their minds to bait him, as a white man. Probably they had heard of the murder, and—being Nationalists, like all schoolboys—regarded it as a victory. They grinned full in Ellis's face as they passed him. They were trying openly to provoke him, and they knew that the law was on their side. Ellis felt his breast swell. The look of their faces, jeering at him like a row of yellow images, was maddening. He stopped short.

'Here! What are you laughing at, you young ticks?'

The boys turned.

'I said what the bloody hell are you laughing at?'

One of the boys answered, insolently—but perhaps his bad English made him seem more insolent than he intended.

'Not your business.'

There was about a second during which Ellis did not know what he was doing. In that second he had hit out with all his strength, and the cane landed, crack! right across the boy's eyes.

G Orwell, pp. 28–30.

Question • Why did Ellis hit the Burmese boy?

A drawing from *The Graphic*, 1902. The artist seems to see nothing wrong with the action of the station-master, indeed, regards it as somewhat amusing. Note: the caption on the drawing is in Hindustani, the everyday language of north India.

Source 7.12 Brutalities experienced by coolies on foreign-owned plantations in Vietnam, *c.* 1928

All the labourers are under the control of the plantation assistant, Mr. V . . ., a Belgian of 23 years of age. The labourers complain of brutal treatment both by Mr. V . . ., who is particularly cruel, and by the overseers under him. . . . Mr. V . . . took [the worker who had attempted to escape], still in handcuffs, and placed him before the rest of the coolies assembled for the roll-call in the central square of the camp. He then ordered the cai [foreman] of Le-Van-Tao's gang, Le-Van-Taon, to hold him by the feet, and another Annamite, who has not been identified as no one dared to denounce him, to take his hands. According to the statements made by Le-Van-Tao and many other witnesses (Tien-Khan, No. 645, Van-Thinh, No. 642, and 16 others), Tao was thus held hanging in the air about 20 centimetres from the ground, his trousers having been removed . . . Mr. V . . . then inflicted on Le-Van-Tao 26 strokes with the lash, breaking the skin in several places and producing wounds which were still suppurating when I examined the laborer . . . .

M J Murray, p. 279.

Question • To what extent is this report an accusation of French colonialism?

Asia's Disillusionment with the West

Increasingly disenchanted with the performance of European colonialism, Asians also became disillusioned with Western civilisation which, not so long before had seemed, in its material splendour, something to be envied and imitated.

One of the first Asians to throw off the cultural cringe and openly criticise aspects of Western society was the Bengali Kayastha philosopher and teacher Narendranath Datta (1863–1902), or, as he became known in adult life, Swami

Vivekananda. Courtesy of his patron, the Maharaja of Mysore, Vivekananda visited the United States in 1893 to attend the first World Parliament of Religions. He was very popular, especially with the ladies, but he was wholly unimpressed by his hosts' wealth and technological progress. As he explained to the Maharaja in a letter from Chicago, where the Parliament was being held, the Americans had achieved much, but at the price of forfeiting their souls (source 7.13).

Another to speak out forcefully on the subject of Western civilisation was M K Gandhi (1869–1948), who was later to lead the Indian National Congress in its struggle against British imperialism. Gandhi had also seen the West at first hand—as a law student in England during the 1890s—and, like Vivekananda, was appalled by its materialism. His essay '*Hind Swaraj*' ('*Free India*') was written in 1909 when he was working as a barrister in South Africa (source 7.14). It conveys the gist of his rather quirky views on money, industrialisation and drink.

But the real turning point in Asian attitudes towards Western civilisation was World War I (1914–18), which showed all too clearly what modern science and technology were capable of. One of the prime movers behind China's Hundred Days of reform in 1898, Liang Qichao (1873–1929) was in his earlier years an ardent moderniser. However, a visit in 1918 to war-ravaged Europe dented his confidence in progress. The West might have conquered nature, he wrote in a newspaper article on his return, but it had not found happiness (source 7.15).

Source 7.13 Swami Vivekananda's impressions of America, 1894

Sri Narayana bless you and yours. Through your Highness' kind help it has been possible for me to come to this country. Since then I have become well-known here, and the hospitable people of this country have supplied all my wants. It is a wonderful country and this is a wonderful nation in many respects. No other nation applies so much machinery in their everyday work as do the people of this country. Everything is machine. Then again, they are only one-twentieth of the whole population of the world. Yet they have fully one-sixth of all the wealth of the world. There is no limit to their wealth and luxuries. Yet everything here is so dear. The wages of labor are the highest in the world; yet the fight between labor and capital is constant.

Nowhere on earth have women so many privileges as in America. They are slowly taking everything into their hands and, strange to say, the number of cultured women is much greater than that of cultured men. Of course, the higher geniuses are mostly from the rank of males. With all the criticism of the Westerners against our caste, they have a worse one—that of money. The almighty dollar, as the Americans say, can do anything here.

No country on earth has so many laws, and in no country are they so little regarded. On the whole our poor Hindu people are infinitely more moral than any of the Westerners. In religion they practice here either hypocrisy or fanaticism. Sober-minded men have become disgusted with their superstitious religions and are looking forward to India for new light.

W T de Bary (Ed), *Indian Tradition*, p. 101.

Question • With which of Vivekananda's criticisms would you **a** agree and **b** disagree?

Source 7.14 M K Gandhi's condemnation of Western civilisation, 1909

Let us first consider what state of things is described by the word 'civilisation'. Its true test lies in the fact that people living in it make bodily welfare the object of life. We will take some examples. The people of Europe today live in better-built houses than they did a hundred years ago. This is considered an emblem of civilisation, and this is also a matter to promote bodily happiness. Formerly, they wore skins, and used spears as their weapons. Now, they wear long trousers, and, for embellishing their bodies, they wear a variety of clothing, and, instead of spears, they carry with them revolvers containing five or more chambers. If people of a certain country, who have hitherto not been in the habit of wearing much clothing, boots, etc., adopt European clothing, they are supposed to have become civilised out of savagery. Formerly, in Europe, people ploughed their lands mainly by manual labour. Now, one man can plough a vast tract by means of steam engines and can thus amass great wealth. This is called a sign of civilisation. Formerly, only a few men wrote valuable books. Now, anybody writes and prints anything he likes and poisons people's minds. Formerly, men travelled in waggons. Now, they fly through the air in trains at the rate of four hundred and more miles per day. This is considered the height of civilisation. It has been stated that, as men progress, they shall be able to travel in airships and reach any part of the world in a few hours. Men will not need the use of their hands and feet. They will press a button, and they will have their clothing by their side. They will press another button, and they will have their newspaper. A third, and a motor-car will be in waiting for them. They will have a variety of delicately dished up food. Everything will be done by machinery. Formerly, when people wanted to fight with one another, they measured between them their bodily strength; now it is possible to take away thousands of lives by one man working behind a gun from a hill. This is civilisation. Formerly, men worked in the open air only as much as they liked. Now thousands of workmen meet together and for the sake of maintenance work in factories or mines. Their condition is worse than that of beasts. They are obliged to work, at the risk of their lives, at most dangerous occupations, for the sake of millionaires.

Formerly, men were made slaves under physical compulsion. Now they are enslaved by temptation of money and of the luxuries that money can buy. There are now diseases of which people never dreamt before, and an army of doctors is engaged in finding out their cures, and so hospitals have increased. This is a test of civilisation. Formerly, special messengers were required and much expense was incurred in order to send letters; today, anyone can abuse his fellow by means of a letter for one penny. True, at the same cost, one can send one's thanks also. Formerly, people had two or three meals consisting of home-made bread and vegetables; now, they require something to eat every two hours so that they have hardly leisure for anything else. What more need I say? All this you can ascertain from several authoritative books. These are all true tests of civilisation. And if anyone speaks to the contrary, know that he is ignorant. This civilisation takes note neither of morality nor of religion. Its votaries calmly state that their business is not to teach religion. Some even consider it to be a superstitious growth. Others put on the cloak of religion, and prate about morality. But, after twenty years' experience, I have come to the conclusion that immorality is often taught in the name of morality. Even a child can understand that in all I have described above there can be no inducement to morality. Civilisation seeks to increase bodily comforts, and it fails miserably even in doing so.

R Iyer (Ed), pp. 213–14.

Question • Is Gandhi rejecting Westernisation, or modernisation, or both?

Source 7.15 Liang Qichao on the bankruptcy of science, 1919

Those who praised the omnipotence [great power] of science had hoped previously that, as soon as science succeeded, the golden age would appear forthwith. Now science is successful indeed; material progress in the West in the last one hundred years has greatly surpassed the achievements of the three thousand years prior to this period. Yet we human beings have not secured happiness; on the contrary, science gives us catastrophes. We are like travelers losing their way in a desert. They see a big black shadow ahead, and desperately run to it, thinking that it may lead them somewhere. But after running a long way, they no longer see the shadow and fall into the slough of despond. What is that shadow? It is this 'Mr Science'. The Europeans have dreamed a vast dream of the omnipotence of science; now they decry its bankruptcy. This is a major turning-point in current world thought.

T-T Chow, p. 328.

Question

- Why was the 'omnipotence of science' widely questioned (both by Asians and Europeans) around 1919?

Renaissance of Eastern Culture

As we have seen, nationalism in Asia began as an idea imported from Europe which took hold among Western-educated élites in the cities. It was fuelled by growing disenchantment with the policies of the colonial regimes, and received added inspiration from the struggles for liberty in Ireland and Russia.

Nevertheless, the early nationalist movement suffered from two major disadvantages. First, its rhetoric was foreign; second, it had a limited appeal to the masses for whom talk of civil rights, and constitutions, and the ending of racial discrimination in selection for the public service, had little meaning. And the support of the mass of the population was vital for the nationalists to remove the Europeans from their positions of power. Obviously, the movement had to be made more **indigenous** and popular. But how? Vivekananda showed the way. By directly challenging the widely accepted idea that the West with its science and technology was a superior civilisation, the Swami stimulated a renewed pride in Asia's traditional values. Interest was especially revived in those religious values which the Westernisers had rejected as superstitious and irrational.

indigenous native to a region

By the turn of the 20th century, India in particular was in the grip of a religious and cultural **renaissance**. The Arya Samaj movement in Punjab and the Ganapati and Shivaji Festivals in Maharashtra were but two manifestations of this renaissance.[2] Seizing their opportunity, politicians like Poona's Bal Gangadhar Tilak (1856–1920) and Bengal's Aurobindo Ghose (1872–1950) used these occasions to convey their message of nationalism to the people in a form which they could understand. Tilak evoked the example of the 17th century Maratha guerilla leader and folk hero Shivaji (source 7.16), and Ghose painted nationalism as a creed

renaissance rebirth; revival of pride and interest in Asian culture sanctified by Bengal's mother-goddess, Durga/Kali (source 7.17).

In India, the medium for the new nationalism was Hinduism. In the Dutch Indies, it was modernist Islam, which gave birth to the Muhammadiyah movement in 1908.

And in Japan it was the cult of Shinto, which, like Hinduism, experienced a patriotic revival around the turn of the century. Okawa Shumei was an ultra right-wing official of the South Manchurian Railway, who figured in several of the attempted military coups of the early 1930s. Source 7.18, from *The Way of Japan and the Japanese* by Shumei, explains how 'Heaven' has chosen Japan to liberate Asia from Western control.

Source 7.16 Bal Gangadhar Tilak on the heroism of Shivaji, 1897

Let us even assume that Shivaji first planned and then executed the murder of Afzal Khan. Was this act of the Maharaja good or bad? This question which has to be considered should not be viewed from the standpoint of the Penal Code or even of the *Smritis* of Manu or Yajñavalkya or from the principles of morality prescribed in Western or Eastern ethical systems. The laws which bind society are for common men like you and me. No one seeks to trace the genealogy of a *rishi* [seer], or to fasten guilt upon a king. Great men are above the common principles of morality. These principles do not reach the place on which great men stand. Did Shivaji commit a sin in killing Afzal Khan? The answer to this question can be found in the Mahabharata itself. Shrimat Krishna preached in the *Gita* that we have a right even to kill our own *guru* and our kinsmen. No blame attaches to any person if he is doing deeds without being actuated by a desire to reap the fruit of his deeds. . . . If thieves enter our house and we have not strength enough in our fists to drive them out, we should without hesitation lock them up and burn them alive. God has not conferred upon the foreigners the grant inscribed on a copper plate to the Kingdom of Hindustan. . . . Do not circumscribe your vision like a frog in a well; get out of the Penal Code, enter into the lofty atmosphere of the *Shrimat Bhagavad Gita* and then consider the actions of great men.

S A Wolpert, pp. 86–7.

Question • Why did the Government of India think this speech was seditious?

Source 7.17 Aurobindo Ghose on religious nationalism, 1907

There is a creed in India today which calls itself Nationalism, a creed which has come to you from Bengal. This is a creed which many of you have accepted when you called yourselves Nationalists. Have you realised, have you yet realised what that means? Have you realised what it is that you have taken in hand? Or is it that you have merely accepted it in the pride of a superior intellectual conviction? You call yourselves Nationalists. What is Nationalism? Nationalism is not a mere political program; Nationalism is a religion that has come from God; Nationalism is a creed which you shall have to live. Let no man dare to call himself a Nationalist if he does so merely with a sort of intellectual pride, thinking that he is more patriotic, thinking that he is something higher than those who do not call themselves by that name. If you are going to be a nationalist, if you are going to assent to this religion of Nationalism, you must do it in the religious spirit. You must remember that you are the instruments of God.

W T de Bary (Ed), *Indian Tradition*, pp. 728–9.

Questions

- Would Aurobindo's definition of nationalism be widely accepted today? If not, why not?

Source 7.18 Okawa Shumei on Japan's historic mission, *c.* 1937

Asia's stubborn efforts to remain faithful to spiritual values, and Europe's honest and rigorous speculative thought, are both worthy of admiration, and both have made miraculous achievements. Yet today it is no longer possible for these two to exist apart from each other. The way of Asia and the way of Europe have both been traveled to the end. World history shows us that these two must be united; when we look at that history up to now we see that this unification is being achieved only through war. Mohammed said that 'Heaven lies in the shadow of the sword,' and I am afraid that a struggle between the Great Powers of the East and the West which will decide their existence is at present, as in the past, absolutely inevitable if a new world is to come about. The words 'East–West struggle,' however, simply state a concept and it does not follow from this that a united Asia will be pitted against a united Europe. Actually there will be one country acting as the champion of Asia and one country acting as the champion of Europe, and it is these who must fight in order that a new world may be realised. It is my belief that Heaven has decided on Japan as its choice for the champion of the East. Has not this been the purpose of our three thousand long years of preparation? It must be said that this is a truly grand and magnificent mission. We must develop a strong spirit of morality in order to carry out this solemn mission, and realise that spirit in the life of the individual and of the nation.

W T de Bary (Ed), *Japanese Tradition*, pp. 795–6.

Question

- How was Okawa's prediction realised during the 1940s?

Overthrowing the Colonial Yoke

By the early part of the 20th century the nationalist movement in Asia had solved the problem of how to mobilise the masses. It was still unsure, however, of how best to achieve its self-appointed task of getting rid of the colonial regimes. In particular, the nationalists were everywhere deeply divided about the effectiveness of violence.

Some people believed that terrorism—especially the killing of high officials—was the quickest and surest way to defeat the Europeans. One such believer was the shadowy Bengali woman 'Madam Cama', who fled India during the crackdown against the anti-Partition agitation of 1905–08,[3] and took up residence, first on the American west coast, and then in Paris. Source 7.19, which comes from her journal *Shabash*, was written shortly after the attempted assassination of the Viceroy Lord Hardinge during a procession to mark the coronation of King George V.

However, others felt that indiscriminate violence was morally wrong and likely to backfire. On the eve of the launching of an India-wide **Non-Co-operation Movement** by the Congress in August 1920, Gandhi used the columns of his newspaper *Navajivan* to exhort his fellow nationalists not to resort to bloodshed (source 7.20).

Non-Co-operation Movement campaign by Gandhi to force British to leave India, whereby Indians refused to be part of or adhere to the administration

Likewise, nationalism in Asia was plagued by disagreements over ideology. Secularists insisted that religion should be kept out of politics, and clashed with fundamentalists, like Aurobindo and Tilak, who believed that politics without religion was bankrupt. At the same time, Communists loyal to Moscow fought with non-Communists who thought that the national struggle should stand on its own feet. For a middle-of-the-road nationalist like Sukarno of Indonesia (1901–70), factional unity often seemed an impossible dream (source 7.21).

Source 7.19 'Madam Cama' in praise of the bomb, 1913

The best reason for the pursuance of the bomb and pistol policy is that there is no more efficacious way of dealing with the situation. There is no better instrument than the bomb. The roar of the bomb represents the voice of the united nations. Who does not understand this? The Madrassi and the Bengali, the Punjabi and the Pathan, the educated and the uneducated—all understand the meaning of the bomb. How are we to convey the message of freedom to the Sikhs, the Gurkhas, and the Pathans in the Indian Army? These people are, in the first place, uneducated, and, secondly, they are confined in cantonments where it is difficult to approach them. But the tyrant himself gave us the opportunity. On the 23rd December, 1912, the Alibaba of the English thieves, the Viceroy of India, mounted an elephant and started in a procession through the streets of Delhi. Both sides of the road were lined with thousands of Indian soldiers. Rajas, Maharajas and hundreds of thousands of men, women and children were assembled to testify to the grandeur of the Government on every side. But the hidden lightning of revolution was also present, ready to demolish the seven-storied tower of the pride of the mischievous. The bomb demonstrated to the Sikhs, Pathans, Gurkhas, Rajas and Maharajas in three seconds that the British kingdom in India was about to come to an end. The bomb in question was a national warning, by beat of drum, to the brave men of India to grid up their loins and come out into the field of battle. It was the voice of the goddess of liberty crying, 'You that are prepared to sacrifice your lives, be up and doing.' Who did not hear that cry?

J C Ker, pp. 130–1.

Questions

- Why does Madam Cama believe that 'there is no better instrument than the bomb'?
- Could bombing attacks such as the one described above actually be counter-productive?

Source 7.20 'Mahatma' Gandhi on the importance of non-violence, 1920

August 1 is are already upon us. All manner of objections are being raised against non-co-operation, the most important being that it is bound to lead to violent disturbances.

It is quite easy to save ourselves from this possible danger. There should be no difficulty at all in preserving peace if there are even a few persons at every place working towards that end. The very first requirement of non-co-operation is to preserve peace. If we do not know how to do this, we have no right to start non-co-operation.

Some people cite the example of the Sinn Fein movement in Ireland and say that non-co-operation and assassinations go on simultaneously there. This is quite true, but Ireland will not get Home Rule that way. There is, moreover, an important difference between us and Ireland. We can very easily gain our object by non-violent non-co-operation and, if violence breaks out, non-co-operation will stop that very moment. Through the method of violence and bloodshed, the people can get nothing in a big country like India; likewise, no power in the world can rule over such a vast country as ours if there is peaceful non-co-operation.

That is why our best success lies in preventing the outbreak of violence. If violence does break out, we ourselves would, and we ought to, rush immediately to the help of the Government to stop it. Breaking out of riots can only mean that we have failed to acquire control over our own people. The second objection to non-co-operation is that the people are not at all ready for it. This really means that the people have no capacity for self-sacrifice.

M K Gandhi, pp. 99–100.

Question • What recent events in Ireland would have been in the minds of Gandhi's readers in 1920?

ahatma' Gandhi and venty-two other *tyagrahis* (demonrators) set out from hmedabad, India, on a 30-mile march in otest at the taxation f salt. This march, in 930, marked the eginning of the ongress-led Civil isobedience ovement, which lasted ntil 1934.

Source 7.21 Sukarno on the problem of reconciling Marxists and Muslims, 1926

Is it possible in a colonial situation for a nationalist movement to join with an Islamic movement, which is part of an international struggle? Can Islam, a religion, co-operate in facing the colonial authorities with nationalism, which is primarily concerned with the nation, and with Marxism, which is based on the philosophy of materialism? . . .

We say with firm conviction: 'Yes, it can be done'. Admittedly nationalism does not concern itself with factions which do not follow in 'the desire to live as one' with the people. It is true that nationalism belittles all factions which do not feel as 'one group, one nation' with the people. And it is true that nationalism fundamentally opposes all forms of action which do not originate from the 'common experience of the people'. But it should be kept in mind that the men who built the Islamic and Marxist movements in our country did have a 'desire to live together as one' with the men who built the nationalist movement. They felt to be 'one group and one nation' with the nationalists, that all groups in our movement: nationalists, Muslims, and Marxists alike, have a history of shared experience behind them, a common fate of being deprived of freedom for hundreds of years . . .

The nationalists who are reluctant to seek contact with Marxists and work together with them show great ignorance of history and of the way the world's political system has evolved. They do not realise that the Marxist movement in Indonesia and Asia generally has the same origins as their own movement. They forget that the objectives of their own movement are often similar to those of the Marxist movement in their country. They forget that to oppose those of their countrymen who are Marxists is to reject comrades in the same struggle and to add to the number of their enemies. They forget or do not understand the significance of the policies of their fellow fighters in other Asian countries, such as the late Dr Sun Yat-Sen, that great nationalist leader who happily and wholeheartedly co-operated with the Marxists, even though he realised that a Marxist organisation of society was still impracticable in China because the necessary conditions did not exist . . .

We are convinced also that we can bring the Muslims and the Marxists together, although the differences of principle between these two are really very great. We are very sad when we recall the blackening of the Indonesian sky several years ago when there was a civil war-like clash, an outbreak of enmity between Marxists and Muslims, when we saw the forces of our movement divided into two factions warring with each other.

This split represents the blackest page in our history. While our movement should have been growing in force, this conflict resulted in the useless dissipation of our strength. It set our movement back by decades.

C L M Penders (Ed), pp. 308–9.

Questions

- Why do Marxists and Muslims generally find it hard to work together?
- Are there any modern cases that you can think of where there has been a successful merging of the two ideologies?

8 Processes of Decolonisation

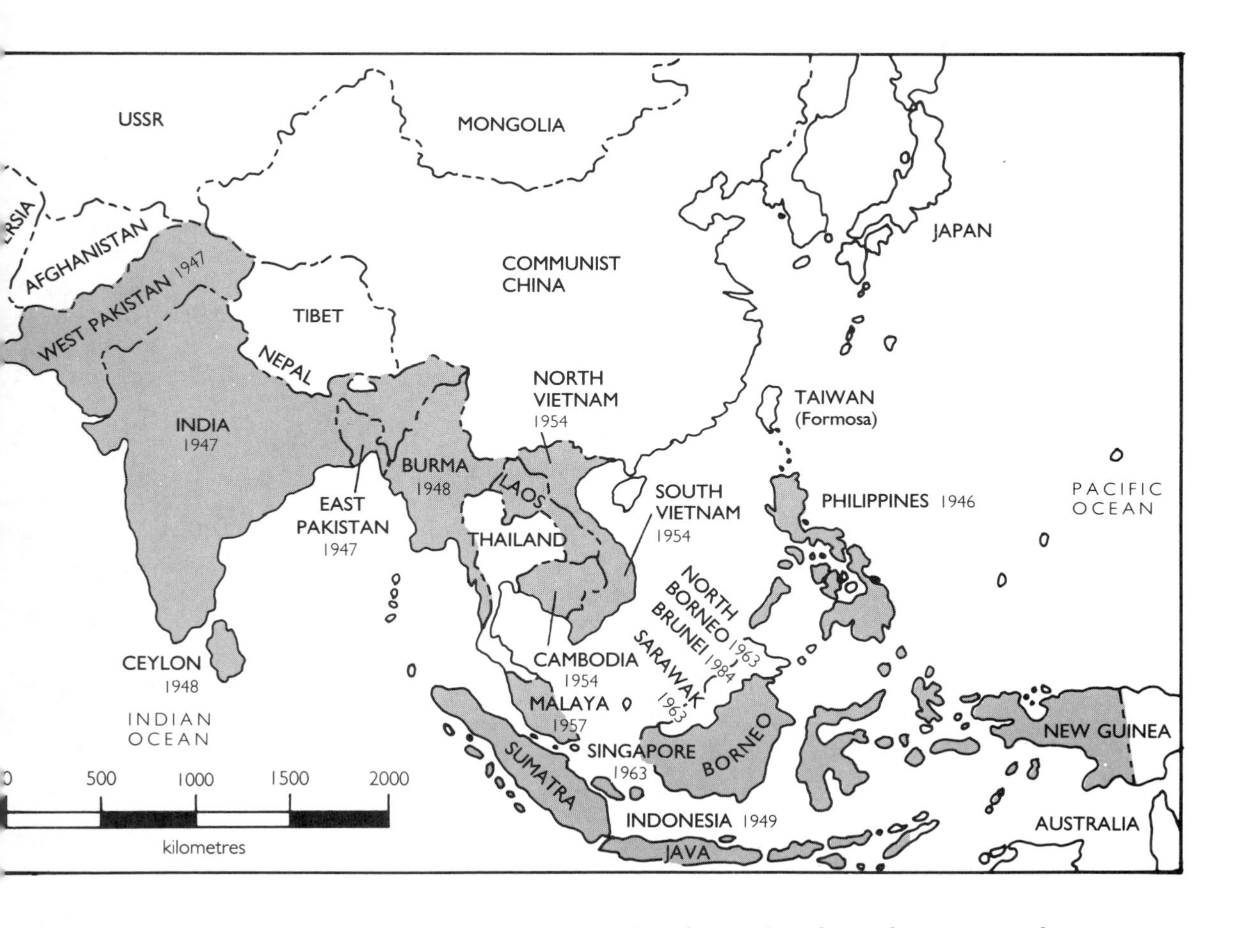

There is no empire [that has been] lost by a free grant of concessions by the rulers to the ruled. . . . Empires are lost by luxury, by being too . . . bureaucratic or overconfident or from other reasons. But an empire has never come to an end by the rulers conceding power to the ruled . . .

Bal Gangadhar Tilak, 1906

EUROPE'S COLONIAL EMPIRES IN ASIA—the fruit of some 300 years of exploration, trade, war and occupation—did not long survive World War II. Indeed, the most important of them collapsed, rather ingloriously, during the period 1946–54. Why did this sudden demise occur? Did the Europeans leave by choice, having fulfilled their 'mission', or were they pushed out by the irresistible force of revolutionary nationalism? This chapter surveys the role of these and other factors in the process of 'decolonisation'.

KEY DATES

- **1909** 'Morley–Minto' reforms establish limited representative government in India.
- **1917** 'Balfour Declaration' promises British support for the Zionist movement campaigning for the setting up of a Jewish homeland in Palestine.
- **1918** German philosopher Oswald Spengler publishes *The Decline of the West.*
- **1927** Congress Against Imperialism and Colonial Oppression meets in Brussels.
- **1930** Unsuccessful rebellion at Yen Bay, Vietnam, led by members of the Viet Nam Quoc Dan Dang (Vietnamese Nationalist Party).
- **1935** United States Congress passes Tydings/McDuffie Act establishing the Commonwealth of the Philippines, and promising full independence in 10 years.
- **1942** Japanese overrun the Philippines, Malaya, Burma and the Dutch East Indies.
- **1943** Under pressure of war, Britain and France renounce their extraterritorial rights in China.
- **1945** Declaration of Independence of the Republic of Vietnam.
- **1954** Under General de Castro, 11 000 French Legionnaires surrender to the Vietnamese forces of General Vo Nguyen Giap at Dien Bien Phu.

ISSUES AND PROBLEMS

Secure in their power, comforted by the conviction that they were doing, if not God's work, then at least good work in Asia, the European colonialists could see no reason ever to leave. Indeed, firmly believing that Asians were incapable of ruling themselves, they could see no easy or morally justifiable way of getting out. As historian Sir John Seeley wrote in 1883, 'to withdraw our

Government from a country [India] which is dependent on it and which we have made incapable of depending upon anything else, would be the most inexcusable of all conceivable crimes'.[1]

trustees guardians, attending to the welfare of native subjects and preparing them for self-government

As time went by this 'illusion of permanence' was modified to the extent that the colonial Powers began to refer to themselves as **trustees** of the natives whom they ruled. During World War I, the British and Americans went as far as putting a time limit on their trusteeship. In 1916 the US Congress passed the Jones Law which promised independence to the Philippines 'as soon as stable government can be established'.[2] In the following year the British Secretary of State, Edwin Montagu, informed Parliament that henceforth, London's India policy would be directed towards 'the increasing association of Indians in every branch of the administration and the gradual development of self-governing institutions'.[3]

Nevertheless, with the possible exception of the United States, the Western Powers were in no hurry to put this timetable into effect. In 1925, a committee on Indian Army reform came up with a proposal which would have seen just half of the officer corps 'Indianised' by the 1980s. And in 1931 a new Governor-General of the East Indies was reported as saying: 'We Dutch have been here for three hundred years; we shall remain for another three hundred. After that we can talk [de-colonisation]'.[4]

communiqué official report or communication

Judging by the **communiqué** issued by the French Government after the Brazzaville Conference of 1944, as late as the 1930s and perhaps as late as the 1940s, decolonisation was not really on the European agenda. Yet, within two decades, almost all the Asian colonies were independent. The first Asian colony to become independent was the Philippines in 1946; a year later, it was the turn of India and Pakistan; in 1948 Ceylon and Burma achieved statehood; in 1949 Indonesia won its freedom; then, after a short interval, followed Indo-China in 1954, Malaya and Singapore in 1957, and Sarawak and North Borneo in 1963. What caused this sudden collapse of colonialism? Was it all due to World War II and the Japanese invasion of 1941–42? Or were the Japanese merely the straw that broke the camel's back? Were the colonial empires already rotten at the core?

Certainly, any explanation of decolonisation in Asia must pay some attention to the War and to the growth of militant nationalisms. But it must also take into account imperial attitudes and policies. In the last resort, the British and the Americans were willing, even glad, to give up their Asian possessions, but the French and the Dutch were not nearly so keen to leave. After the War both tried, unsuccessfully, to reclaim their colonies and to return to the *status quo*. Why were the latter Powers so unbending? Were their Asian colonies relatively speaking more valuable or prestigious? Or were their politicians simply too short-sighted to realise that their era was over?

THE DEBATE

Historians of the British empire, especially, used to peddle the thesis that decolonisation in Asia was voluntary and the result of a long-term plan. A typical example is Sir Percival Griffiths' summation in his book, *The British Impact on India*:

> Other ruling powers have abdicated after defeat in war or as a result of successful insurrection, but it was left for Britain to surrender her authority of set purpose and as part of a process of [d]evolution which had been operating for some decades. She acted under no external compulsion. In 1942, although at the nadir of her military fortunes, she had unmistakeably crushed a carefully prepared rebellion in India; and in 1945 she and her allies had completely defeated their Western and Eastern enemies. . . . It was at that moment of supreme power that Britain transferred full authority to India and Pakistan and demonstrated, once and for all, her belief that self-government is the only proper end of the colonial system.[5]

More recently, however, some British students of empire have come round to the view that more cynical calculations were involved. For instance, B R Tomlinson has argued convincingly that the British quit India because it was no longer an economic asset. Ronald Robinson suggested, however, that colonial regimes gave up only when 'They had run out of collaborators'.[6] But even this theory accepts that the final decision about when to leave rested with the colonial authorities. By contrast, most Asian writers on this period insist that the Europeans' departure was entirely involuntary, that they were forced out by the power of revolutionary nationalism. Within this larger debate, argument also rages about the importance of World War II. 'More than any other nation', writes American War historian Edwin P Hoyt, 'Japan caused the end of colonialism [in Asia]'.[7] Other historians, however, believe that the die was well and truly cast before 1939, and that the Japanese invasion merely accelerated an already fatal process of structural decline.

Europeans' Illusion of Permanence

Well into the 20th century, colonial administrators in Asia and imperial statesmen in Europe were similarly convinced that their **tenure** of political power in that part of the world still had centuries to run and that they had nothing to fear from nationalist unrest. Indeed, most of them were unable to comprehend the genuine feelings of outrage and betrayal that motivated the nationalists in the colonies. And they dismissed the demands of parties like the Indian National Congress as the selfish outpourings of an unrepresentative minority. As the Tory Viceroy Lord Minto wrote to the Liberal Secretary of State for India, John

> **tenure** condition under which property is held; period of holding or possession

Morley, in November 1906, at the height of the anti-Partition agitation in Bengal: 'There is no popular movement from below' (source 8.1).

Moreover the Europeans had no intention of leaving the colonies whatever the nationalists did or said. The colonies were too valuable, and their peoples had not yet learned to rule themselves. Take away the firm hand of Britain, declared Brigadier-General Colomb, speaking from years of experience with the Indian Army, and you would end up with anarchy (source 8.2). As late as the 1930s, **die-hard** writer Nesta Webster probably spoke for the majority of her countrymen when she asserted that the whites were in Asia and Africa 'by right of conquest, and they intend to stay there'.[8]

'die-hards' British Conservative Party members opposing substantial political reform in India

Source 8.1 Lord Minto: Why Asian political movements are not truly nationalist, 1906

What is going on in India is altogether peculiar in comparison with other revolutions. Gambetta and Clemenceau, and before them Cavour, Garibaldi, and Mazzini, were fighting for what they believed to be the liberties of the people and had the support of a great majority of their fellow countrymen. I have always thought the regeneration of Italy a very fine story, though it was led by extremists who were not over scrupulous; but here the position is entirely different. There is no popular movement from below. The movement such as it is, is impelled by the leaders of a class very small indeed in comparison to the population of India, who, if by some miracle they obtained the reins of Government, are totally incapable of ruling and would not for an instant be tolerated by the people of India as a whole.

C H Philips (Ed), p. 78

Question

- Minto draws a number of distinctions between nationalist movements in Europe and in India. Are any of his points, in your view, valid?

Source 8.2 G H Colomb on 'Britain Without India', 1925

The creeds and classes of all time in India are still there, controlled by the gentle hand of British power, but ever ready to fly at each other's throats the moment opportunity offers. With the removal of British protection at this early stage, dissensions would lead to quarrels, quarrels to bloodshed; Civil and religious war would break out as a result of the first clash between Hindus and Mahomedans [Muslims], on the occasion of one or other of their respective festivals. The spark, once allowed to glow, would soon be fanned into flame, and uncontrolled, would spread like lightning from town to town. Wild destruction, orgy, arson and butchery would promptly follow. In those parts of the country where Hindus predominate, the latter would hold the upper hand. Similarly. in other districts, Mahomedans would have control. Leaders from both factions would arise locally, only to find themselves outclassed and outpowered by superior bands of the opposite creed. Bridges, buildings and railroads would be demolished and torn up, and all means of communication injured to the utmost extent. There would be no thought then of British rule. The primary consideration would be creed against creed and war to the knife. There would be fanatical leaders everywhere, one vying with another in cruelty.

G H Colomb, p. 893.

Question

- Did clashes like the religious civil war envisaged by Colomb actually come to pass?

Doctrine of Trusteeship

Nevertheless, it would be unfair to suggest that the colonial Powers were all of like mind on the question of Asian self-government. William Howard Taft (1857–1939) was the Philippines' first civilian Governor. In 1912 he reminded President Woodrow Wilson that the United States had from the first indicated that its dominion in the Philippine Islands would last only until such time as the Filipino people showed themselves 'reasonably fit' to govern (source 8.3).

Even the stiffly conservative British Raj was led, under pressure of war, to relax its iron grip on India, introducing a system of elections and appointing an Indian to the Viceroy's Executive Council. Indeed, in August 1917 the British went further, promising to continue the development of self-governing institutions with a view to building in India a system of **responsible government**. For decades the British had rejected any suggestion that Western-educated Indians were capable of self-rule However, as the Joint Report of the Viceroy Lord Chelmsford and the Secretary of State, Edwin Montagu makes clear, the British now conceded that they were indeed deserving of their sympathy and encouragement (source 8.4).

responsible government form of government where executive is formed from the majority party in the elected legislature

Yet not all political changes were introduced in a spirit of liberalism. Montagu's desire to allow Indians to participate more actively in the running of their country seems to have been quite genuine. However, Chelmsford's officials supported the reforms because they appeared to provide a new means of dividing and ruling the opposition. Likewise, the officialdom in the Dutch East Indies was persuaded to support a limited program of Indonesianisation mainly because locals were cheaper to employ than Europeans (source 8.5).

Source 8.3 America's aim in the Philippines, 1912

. . . the national policy is to govern the Philippine Islands for the benefit and welfare and uplifting of the people of the Islands and gradually to extend to them, as soon as they shall show themselves fit to exercise it, a greater measure of popular self-government. . . . What should be emphasised in the statement of our national policy is that we wish to prepare the Filipinos for *popular* self-government. This is plain from Mr. McKinley's letter of instructions and all of his utterances. It was not at all within his purpose or that of Congress which made his letter a part of the law of the land that we were merely to await the organisation of a Philippine oligarchy or aristocracy competent to administer the government and then turn the Islands over to it. . . . Another logical deduction from the main proposition is that when the Filipino people as a whole show themselves reasonably fit to conduct a popular self-government, maintaining law and order and offering equal protection of the laws and civil rights to rich and poor, and desire complete independence of the United States, they shall be given it.

R von Albertini, *Decolonization* p. 475.

Question • Is Taft's suggestion that President McKinley had always intended to prepare the Filipinos for self-government confirmed by source 2.20?

Punch cartoon on the Montagu-Chelmsford reforms, 1919. Given that 'India' is pictured as a tailor's dummy, what do you think *Punch* is saying here about the reform process?

Source 8.4 The Montagu–Chelmsford report on Indian constitutional reform, 1918

In estimating the politically-minded portion of the people of India we should not go either to census reports on the one hand, or to political literature on the other. It is one of the most difficult portions of our task to see them in their right relation to the rest of the country. Our obligations to them are plain for they are intellectually our children. They have imbibed ideas which we ourselves have set before them and we ought to reckon it to their credit. The present intellectual and moral stir in India is no reproach, but rather a tribute to our work. The *Raj* would have been a mechanical and iron thing if the spirit of India had not responded to it. We must remember, too, that the educated Indian has come to the front by hard work; he has seized the education which we offered him because he first saw its advantages; and it is he who has advocated and worked for political progress. All this stands to his credit.

For 30 years he has developed in his Congress and latterly in the Muslim League, free popular convocations [assemblies] which express his ideals. We owe him sympathy because he has conceived and pursued the idea of managing his own affairs, an aim which no Englishman can fail to respect. He has made a skilful, and on the whole a moderate, use of the opportunities which we have given him in the legislative councils of influencing Government and effecting the course of public business, and of recent years he has by speeches and in the press done much to spread the idea of a united and self-respecting India among thousands who had no such conception in their minds.

E Montagu and Lord Chelmsford, para. 139.

Question • Why would T B Macaulay (source 2.12) have approved of this repor

Source 8.5 J W T Cohen-Stuart on the Indonesianisation of the Dutch Colonial Services, 1907

It is of great importance for the continuation of our rule that we should bind the people to us, partly by letting them participate as much as possible in government administration and partly by letting them feel our rule as little as possible. Both these objectives can be achieved most effectively if we restrict ourselves as much as possible to a supervising role and leave the task of governing wherever possible to the natives themselves. In this way also the mistakes that are made will not be blamed so much on us as on their own countrymen . . . The appointment of native officials instead of European ones wherever feasible will also greatly decrease the burden on the budget . . . A European type of administration is too expensive for a country like the Indies . . . Obviously a rich country like the Netherlands with a population six times smaller can easily afford a larger budget than the Indies, which is a poor country. But a serious inconsistency has crept into the Indies budget, which has to provide for a European administration that is twice as expensive as that in the Netherlands. The first thing to be done is to change this situation, if we want to be in a position to provide properly for the unmet needs of the Indies. And this can only be done by gradually replacing European officials by native ones, who will be paid at a lower rate based on the lower standard of living of the people . . .

C L M Penders (Ed), p. 165.

Question • Is Cohen-Stuart suggesting that the Dutch should embark on a program of decolonisation?

Were the Colonies a Declining Asset?

The popular view in Europe was that colonies were assets which made money for and added to the international standing of the metropolis. In fact, as chapter 9 will show, colonies were rather costly to run.

As the 20th century advanced, the costs of colonial ownership went up sharply as the Japanese and other foreign competitors began to supply more of the colonies' imports and as political unrest became more widespread. For example, Britain's share of India's imports went down overall from nearly 76 per cent in 1913–14 to about 45 per cent in 1938–39 (source 8.6).

Another important change, which occurred in the early months of World War II, was the signature of a new defence agreement between Britain and India. Previously, the entire cost of the 200 000 strong Indian Army (England's own 'barrack in the Eastern seas', as it was once smugly described by Lord Salisbury) had fallen on the shoulders of the Indian taxpayer, even when, as during the Boxer rebellion of 1900, it was used to defend imperial rather than Indian interests. Under the 1940 agreement, the cost was to be shared, with the Home Government paying for all campaigning done outside the subcontinent. As a result, India quickly cleared its long-standing debt to Britain and began to accumulate considerable sterling credits. As the Chancellor of the Exchequer (Treasury) Kingsley Wood, explained to the Cabinet in a memorandum of 1942 (source 8.7), the implications of this financial turnabout for Britain's **solvency** were grim.

solvency having sufficient money to meet all financial obligations

Although still valuable strategically and diplomatically, India, and to a lesser extent other Asian colonies, were no longer the economic assets that they had once been.

Source 8.6 Percentage share of British goods in India's imports, 1913–38

	1913–14	*1928–9*	*1938–9*
Cotton piece-goods	94	79	32
Iron and steel	78	56	50
Other metal manufacturers	46	34	34
Hardware and cutlery	56	26	29
Electrical machinery	79	66	57
General machinery	92	76	57
Railway locomotives and carriages	95	88	61
Motor vehicles	66	15	30
Chemicals	75	59	57

B R Tomlinson, p. 47.

Question

- Can you find figures which show the destination of India's exports during the same period?

Source 8.7 Memorandum by the Chancellor of the Exchequer on British war debts to India, 1942

3. The War Financial Settlement of April 1940 provides that India recovers from us all her defence expenditure except her normal pre-war budget (adjusted for rises in prices), measures taken in India for India's local defence, and a share of measures 'taken jointly in the interests of Indian Defence and of H.M.G.' The underlying assumption that the War is not India's affair except so far as the local defence of India is concerned, may have been a tenable point of view in 1939 and 1940. It is no longer so in 1942.

4. The result of this arrangement (in conjunction with India's favourable balance of visible trade) has been that we have paid India very large sums. Hitherto, India has been able to use these payments to extinguish her indebtedness to us in respect of pre-war sterling

loans. These loans have now been entirely repatriated, except for Railway Debt to the amount of some £50 millions. From now onwards India is likely to increase her sterling balances, or, in other words, we are likely to become heavily indebted to India. By April 1943 India's sterling balances may amount to between £400 and £450 millions (of which some £250 millions represent backing for the note issue) . . .

5. The arrangement of April, 1940, was not accepted by the Chancellor of the Exchequer as necessarily final. The whole position has been completely changed by the Japanese entry into the war and the resulting threat to India, and by the adoption of the principle of Reciprocal Aid. I feel bound to submit to my colleagues that the existing arrangement should now be radically revised.

6. The acquisition of very large sterling balances by India would give rise to grave problems, both financial and political. Our greatest economic difficulty after the war will be to preserve equilibrium in our balance of payments with the rest of the world—i.e. to pay for the imports needed to maintain a tolerable standard of living for our people. We cannot afford to start with a handicap of debt amounting to several hundred millions of pounds. It would be intolerable that this should be our reward for successfully defending India from being overrun by Japan or Germany.

7. India would certainly demand that these balances should be converted, wholly or partly, into gold or dollars. This would be impossible and political friction would result: we should be told that a rich country had repudiated a debt to a very poor country. There would be no satisfactory method of dealing with very large balances after the war. It is essential that they should not be allowed to accumulate. . . .

10. I have stated our point of view to Sir Jeremy Raisman, the Finance Member [of the Government of India], who has come to this country on a short visit to discuss this question. He is, as I gladly recognise, anxious to give any help he can within the limits of the existing financial arrangement. But the help which he can give within these limits is very little (not more than £40 millions non-recurrent and £7½ millions recurrent) and does not advance us towards a solution of the problem. He maintains that the existing War Financial Settlement cannot be altered. Any alteration would, he contends, be viewed by Congress and even moderate public opinion with intense and bitter suspicion, and would raise counter-demands that we should pay in dollars or gold, instead of in sterling. India cannot, he says, be coerced into a war effort free of charge, and sterling is, therefore, bound to accumulate so long as India remains in the war.

Public Record Office: CAB 66/26, pp. 22–3.

Question • How much did Britain owe India by 1945?

Revolutionary Nationalism

In the 20th century, Asian colonies became not only relatively more expensive, but also much more troublesome as nationalists grew more militant and better organised. Despite Gandhi's message of non-violence, acts of political terrorism by the extremist wing of the Indian nationalist movement mounted steadily. Terrorism peaked during the depression of the early 1930s, which brought hard times to India as well as the industrialised world (source 8.8).

In Java, heartland of the Dutch East Indies, activists belonging to the PKI (Partai Komunis Indonesia) made common cause with fundamentalist Muslims to launch a massive revolt against Dutch rule in 1926–27. Taken from the report of the commission appointed by the Batavia government to investigate the 'PKI disturbances', source 8.9 reveals how this unlikely union was forged.

Source 8.10 reproduces one of the inflammatory tracts that were distributed throughout the countryside during this period. Signed 'H T Damste' (a Dutch name and therefore almost certainly an alias), the document incites the Muslim faithful to engage in *jehad* (holy war) against the Dutch, promising those who died in the cause instant union with God in Paradise.

Until World War II, the colonial regimes did not bother to control nationalist dissent. Nevertheless, each successive demonstration or insurrection stretched the colonialists' resources more thinly, and eroded further their claims to rule with the support of the majority of the people.

Source 8.8 'Political crime' in India, 1906–36

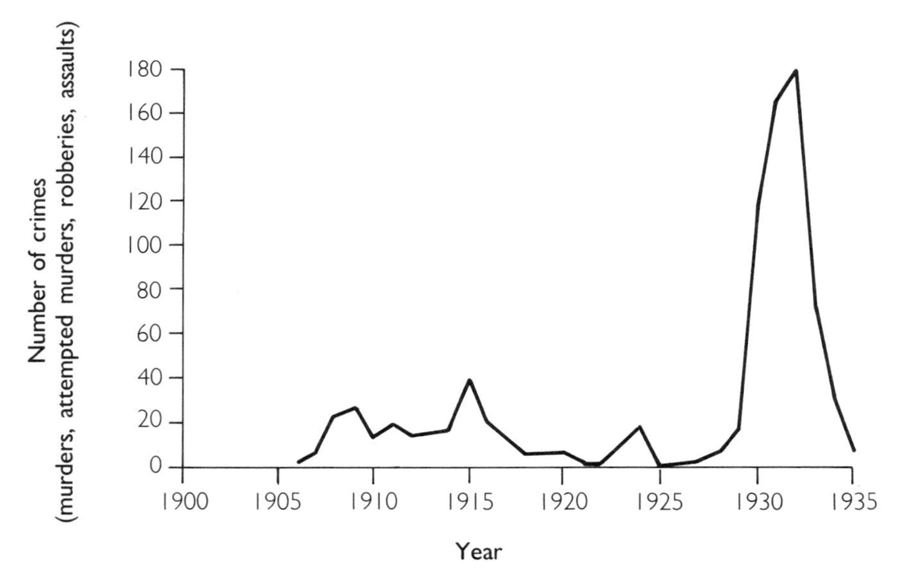

Adapted from J C Ker and H W Hale.

Question

- Why was there a sudden increase in the incidence of 'political crimes' in the early 1930s?

Source 8.9 Report on the Communist insurrection in Banten, Indonesia, 1927

The Promised Utopia

The Communists showed great skill and keen insight in the way in which they spread expectations of the success of the rebellion and promises of a Utopia . . .

The descendants of the sultan and the other

title-bearers were promised the establishment of a new sultanate and 'their own sultan'; this state was represented as an Islamic state to the religious orthodox.

The followers of the religious leaders who were preparing for the rebellion were enticed with the prospects of the glories of paradise, the reward which would await them as warriors victorious in Allah's name, or as martyrs who have died for his cause . . .

[And] everyone was led to expect the blessings of cheap rice or free rice and free transport in cars and trains, etc. But nothing much was said about distribution of property belonging to the wealthy because an attempt was made to get the wealthy to join also.

Side by side with the illusions of fortune for those who would rebel were of course the threats for those who would not. They would not partake in the advantages of Utopia; on the contrary they would be oppressed: their property would be confiscated for the founders of the new community.

The Possibility of Success

Even more important than the notions formed about a Utopia were those concerning the possibility of the success of the rebellion. The crux of the whole problem is that the Bantamese do not like rule, whether it be Dutch or other. And thus the main feature of the Communist action apparently consisted of impressing upon the minds of the population the possibility of rebellion succeeding. To put it even more forcibly, the Communists convinced the people that a rebellion would arise and that the pergerakan was strong and powerful, irresistable and inevitable.

Thousands of members were prepared, part of the police was on the side of the Communists, together with the majority of 'the soldiers of Tjimahi' and 'Batavia,' likewise various prominent gurus, among them the Kiai of Tjaringin. This even made an impression on members of the constabulary. What will you be able to do with sixty men if we turn up with two hundred thousand? The rebels in Menes and Labuan were instructed that after the first rebellious actions 'the soldiers of Tjimahi' would arrive and decide what was to be done with the officials who had not yet been murdered . . .

Religion

Religion is, generally speaking, of such paramount importance in Bantam that a separate study of its influence on the movement cannot be considered out of place here . . .

Whereas other areas endeavored to liberate themselves of the foreign tyrant in order to attain the promised Utopia, many of the Bantamese wished to free themselves of the infidel tyrant in order to pay greater honor to Islam . . .

In North Bantam it became apparent shortly after the arrests made in August that several people had fasted rigorously as the last stage of the ritual which had to be followed in order to become invulnerable. This also proved to be the case in Tjidasari, near Pandeglang, while the same thing occurred among members of the rebel groups in Menes and Labuan, as has been learned from the interrogations held after the riots. Members of these groups were even sent to Hadji Dulhadi of Banko by their leaders. The latter received payment for training invulnerable warriors, who proved their ability by dancing on chopping knives while some of them allowed their invulnerability to be put to the test by gunpowder and parangs—without success, however. All the gurus who taught the doctrine of invulnerability played an important personal part in the rebellion. Hadji Selah of Balagendon (North Bantam), whose father and grandfather were killed in the rebellions against the government in 1888 and 1850, was in his turn killed in Tjaringin.

H Benda and R T McVey (Eds), pp. 43–5

Question • What clues can you find in this document as to the reasons for the failure of the PKI revolt?

Source 8.10 A Muslim revolutionary tract distributed in Java, 1928

Be very devoted and do not worry! You will get God's help. 'Paradise is in the shadow of the swords', this tradition is also very well known . . . This is a road sign to the very beautiful Paradise, which is a haven of comforts. This is the order of things in the afterlife, which is different from things in this world. 'The eternal life is to be found in the drinking of the cup of death.' God gives eternal life after death, oh brothers. Listen again to the following verse which I will read to you about the blessings and pleasures which He will readily give you: 'Oh, believers, shall I teach you the means to escape your deep suffering? Believe in God and in His prophet, fight under the banner of the faith, liberally sacrifice your life and goods; that is your road to happiness, if you want to know. God will forgive your sins. He will lead you into the gardens where streams are flowing. You will enter into the exquisite Garden of Eden where you will enjoy the greatest happiness.'

The easiest way for the faithful is, according to the directions of God, the great Lord, to do battle in the path of Allah . . . then all sins will be forgiven and wiped out, even if they are as numerous as the foam in the surf on the beach, and he will grant the Paradise of Eden, a region of feast and pleasures. There one will get everything that is desired immediately. The Lord gives special pleasures to those who served the cause of the holy war. He will give them heaven, paradise, which glows with an unforgettable light! There will be seventy heavenly nymphs and in addition to that, girl servants. God's reward being so great, is it proper to be slow in taking up battle against the unbelievers?

C L M Penders (Ed), p. 208.

Question

- Why were the Dutch authorities particularly alarmed at the prospect of Muslim fundamentalists declaring a *jehad* against their rule?

Importance of World War II

The role of World War II in the process of decolonisation in Asia has been a subject of much debate. While some historians would say that the colonial regimes were already fatally weakened by 1939 when the War broke out, almost all would concede that the following factors speeded up the transfer of power in Asia: (i) the enormously draining effect of the conflict on the resources of the metropolitan Powers; (ii) the humiliating and disrupting impact of the Japanese invasion of 1941–42; and (iii) the need, faced in particular by Britain (India was still free but under siege), to win the support of colonial peoples for the war effort.

Wartime US Secretary of State, Cordell Hull.

Here, we focus mainly on the last of these factors. As we have seen, the United States had always taken a more liberal approach to the colonial issue than the other Powers. As the War moved towards its conclusion they tried to persuade their partners in the Grand Alliance—Britain, Free France and the Netherlands—to follow the example they had set in the Philippines and set a date for independence. In the words of wartime Secretary of State Cordell Hull (1871–1955): 'We believed that the time had come when all parent countries should begin to plan and prepare for the self-government of these peoples'.

Generalissimo commander of combined military forces

China's wartime leader Generalissimo Chiang Kai-shek photographed in 1953, after he had fled to Taiwan.

Likewise, pressure for decolonisation came from the Nationalist Chinese government of **Generalissimo** Chiang Kai-shek. As an Asian leader, Chiang was naturally sympathetic to the anti-colonial struggle, and since China's military support was vital to the Allies' chances of defeating Japan, his views were heard patiently in Washington. In particular, Chiang received assurances from President Roosevelt that he would take up the issue of Indian independence personally with Prime Minister Churchill. As Madam Chiang Kai-shek jubilantly informed the Congress leader Jawaharlal Nehru in a letter of March 1942, the support of the American President would guarantee an early transfer of power in the Subcontinent (source 8.12).

In the event it was not quite that easy. As Cordell Hull admits in his memoirs, it was difficult to get the colonial Powers to change their attitudes (source 8.11). For instance, Churchill virtually told Roosevelt to mind his own business.

Ultimately, however, the Allies were cornered by their own war propaganda. Having declared, in the Atlantic Charter of 1941 and later in the United Nations Charter, that they were fighting to liberate the world from tyranny, they could hardly justify reimposing another form of tyranny on Asia. By 1945, oppressed peoples everywhere were looking for deliverance, and their claims could not be refused easily (source 8.13).

Source 8.11 The Hull–Roosevelt approach to post-war decolonisation

Concerning the vast area of the Southwest Pacific, my associates and I had been doing considerable thinking and, along with the President, had arrived at certain conclusions during my last years in office. This area embraced such important territories as the Dutch East Indies and the Philippines, and could be taken to include Malaya and French-Indo-China.

These enormous lands entered into the intensive discussions we had been holding on the subject of dependent peoples. Without being specifically mentioned, they were included in the projects I had presented to the British and Russians under that heading.

We believed that the time had come when all parent countries should begin to plan and prepare for the self-government of these peoples, to be given them when they were ready for and worthy of it. Before us we always had the example of the Philippines, whom the United States had been preparing for independence almost since the day of our acquisition of the islands, and for whom an independence date had been formally set by national legislation in the Tydings-McDuffie Act of 1934.

The President was in thorough agreement with our proposals. He himself entertained strong views on independence for French Indo-China. That French dependency stuck in his mind as having been the springboard for the Japanese attack on the Philippines, Malaya, and the Dutch East Indies. He could not but remember the devious conduct of the Vichy Government in granting Japan the right to station troops there, without any consultation with us but with an effort to make the world believe we approved.

From time to time the President had stated forthrightly to me and to others his view that French Indo-China should be placed under international trusteeship shortly after the end of the war, with a view to its receiving full independence as soon as possible. . . .

He then went on: 'As a matter of interest, I am wholeheartedly supported in this view by Generalissimo Chiang Kai-shek and by Marshal Stalin. I see no reason to play in with the British Foreign Office in this matter. The only reason they seem to oppose it is that they fear the effect it would have on their own possessions and those of the Dutch. They have never liked the idea of trusteeship because it is, in some instances, aimed at future independence. This is true in the case of Indo-China' . . .

Our prime difficulty generally with regard to Asiatic colonial possessions, of course, was to induce the colonial Powers—principally Britain, France, and The Netherlands—to adopt our ideas with regard to dependent peoples. Britain had refused to go along with us on the idea of eventual independence for her colonies, believing instead that they should in time achieve self-government within the Empire. We had frequent conversations with these parent countries, but we could not press them too far with regard to the Southwest Pacific in view of the fact that we were seeking the closest possible cooperation with them in Europe. We could not alienate them in the Orient and expect to work with them in Europe.

At no time did we press Britain, France, or The Netherlands for an immediate grant of self-government to their colonies. Our thought was that it would come after an adequate period of years, short or long depending on the state of development of respective colonial peoples, during which these peoples would be trained to govern themselves. Our cause was harmed, not helped, by some vociferous persons in the United States, including Vice President Wallace, who argued for an immediate grant of independence or for the total separation of colonies from their mother countries.

C Hull, pp. 1595–9.

Question • Why was the USA more friendly towards the role of decolonisation in Asia than Britain or France?

Source 8.12 Chiang Kai-shek, Roosevelt and India, 1942

The Generalissimo and I . . . feel that we owe it to our Indian friends to speak the truth as we see it, although as we were guests of the British government, politeness constrains us from openly criticising the assertion that real power cannot be given to India because of the lack of unity among her people etc. I saw in the papers today that the London *Chronicle* made quite a case of this, and was I furious. . . .

The Generalissimo has been telegraphing Roosevelt on Indian conditions. Our latest news from him is this: Roosevelt wired that at the Peace Conference the representative from India should be chosen by *Congress*, and represent real national India. . . . He thinks that a solution of the Indian problem might be found in dividing India into two, namely Moslem and Hindu. Both the Generalissimo and I wired to my brother T V. [Soong] that the second premise is entirely wrong, and should not be considered for one single second. India is as indivisible as China. The fact that there are religious differences amongst her people does not mean that politically they cannot agree if given the opportunity to settle their diversity of views uninterfered with and unabetted by a third party.

D Norman (Ed), pp. 78–9.

Question • Why was Chiang Kai-shek in a good position to put pressure on Washington?

Source 8.13 A petition from 540 citizens of Makassar, Indonesia, to the 'allied nations', 1945

We, rajah's, adat-chiefs, higher officials and leaders of religious, political and social groups in Selebes:

Hearing the proclamation of the Republic of Indonesia for the whole territory of Indonesia;

Regarding the present condition and feeling of the Indonesians in Selebes, demanding that Selebes also should be acknowledged as a part of the Republic of Indonesia;

CONSIDERING

that according to the Atlantic Charter the principles of which are adopted by the conferences of the Allied Nations held later on in Teheran, Yalta and San Francisco and furthermore confirmed by the 12 articles of the President of the United States of America, broadcasted after the end of the late World-war namely on the 28-th of October 1945, the Allied Nations will respect the right of each nation to choose the form and government it desires and want the return of sovereignty and self-government to those who are violently deprived thereof;

that based upon this, the independence of Indonesia should be acknowledged, in the first place because the Indonesians have expressed their desire to establish an independent Republic in behalf of which an own government and administration are already formed; in the second place because the Indonesian people of several regions are successively deprived of their sovereignty and self-government by the Dutch by way of war of conquest and by use of the 'divide and rule'-policy (divide et impera);

that moreover, the Atlantic Charter aims at full cooperation between all nations in the domain of economy with the object to guarantee the improvement of labor conditions, economic progress and social security for all, which can only be realised in a fully independent country —which means: in Indonesia in an independent Indonesia—where all nations without any exception should have the same right of economic admission;

that in reality independence is the right of all nations and therefore oppression must be wiped out of this world because it is against justice and humanity;

that the fight for an independent Indonesia— in which the Indonesians have already shown ability and readiness for sacrifice—has in view to lead the Indonesians to an independent, united, sovereign, righteous and prosperous Indonesia, with an own government which strives after protection of the whole population and territory of Indonesia, advancement of general prosperity, increase of the standard of life and participation in a world-order in which the society is free from want and fear, and based upon an everlasting peace and social justice;

that the Republic of Indonesia stand on a democratic constitution, based upon God's law, a righteous and cultured mankind, a unity of Indonesia, and a society guided by the imprudence of a representation body and by the realisation of social justice for the whole population of Indonesia;

DECLARE

that we, together with the people under our administration and/or our followers, want an independent and democratic Republic of Indonesia which stands on justice and equality of rights where everyone who respects the laws of the country can have a place;

that therefore we want to be an inseparable part of the unity of the Republic of Indonesia.

REQUEST

the Allied Nations to acknowledge the Rupublic of Indonesia.-

In the firm conviction that the Allied Nations in their decisions will be guided by the principles of justice and humanity, which are the essential elements of a real democracy,

H Feith (Ed), pp. 49–50.

Question • Did the tactic of appealing to international opinion help the Indonesians to win their independence?

The Final Curtain Falls

devolution
delegation of work or power by governing body to other bodies appointed by and responsible to it

While Winston Churchill (1874–1965) remained at the helm, the British Government stoutly resisted demands from the nationalists for an early **devolution** of power. 'I have not become His Majesty's First Minister', he told the House of Commons in 1944, 'to preside over the dissolution of the British Empire'.[9]

However, others closer to the action, such as the severe but sensible Viceroy Lord Wavell (1883–1950), recognised the inevitable. In a lengthy and uncharacteristically passionate letter in October 1944, Wavell attempted unsuccessfully to persuade Churchill that Britain's own interests would be served best by accommodating the Congress. In source 8.14 he is shown to advocate an immediate response to the Indian problem. The British Labour Party was also sympathetic to Congress's demands. When Churchill's Conservative Party was rejected by the voters in May 1945, Clement Attlee ordered Wavell to reopen negotiations with the nationalists.

By 1945, then, it was no longer a question of 'if' but 'when'. British power in the subcontinent visibly crumbled in the face of rising civil disobedience and bloody riots between Hindus and Muslims agitating for a separate state of Pakistan. It seemed to the men who were present absolutely vital that power be handed over quickly, before the Raj really was driven out. A joint telegram to London from the Viceroy and the Secretary of State, who had flown out to India to assess the situation, shows how grim things had become (source 8.15).

Unfortunately, though, the British example was not followed by the French, Dutch and Portuguese. Instead of bowing to the inevitable, these colonial Powers became more stubborn. The French made much about a new deal in which the 'former' colonies would be 'associated' as 'autonomous' entities within an over-arching federation—the French Union. But as the debate in the Chamber of Deputies on the future of Indo-China makes clear, this was simply an elaborate, deceptive cloak for continuing control and exploitation (source 8.16).

Source 8.14 Arguments for early Indian independence, 1944

I will begin by saying that my primary reason for writing is that I feel very strongly that the future of India is the problem on which the British Commonwealth and the British reputation will stand or fall in the post-war period. To my mind, our strategic security, our name in the world for statesmanship and fairdealing and much of our economic well-being will depend on the settlement we make in India. Our prestige and prospects in Burma, Malaya, China and the Far East generally are entirely subject to what happens in India. If we can secure India as a friendly partner in the British Commonwealth our predominant influence in these countries will, I think, be assured; with a lost and hostile India, we are likely to be reduced in the East to the position of commercial bag-men.

And yet I am bound to say that after a year's experience in my present office I feel that the vital problems of India are being treated by His Majesty's Government with neglect, even sometimes with hostility and contempt. I entirely admit the difficulty of the problems, I know the vital preoccupations of the European war. I agree in the main with what I think is your conviction, that in a mistaken view of Indian conditions and in an entirely misplaced sentimental liberalism we took the wrong turn with India 25 or 30 years ago; but we cannot put back the clock and must deal with existing conditions and pledges; and I am clear that our present attitude is aggravating the mischief . . .

When we should make any fresh move is a difficult problem. I am quite clear that it should be made some considerable time before the end of the Japanese war. When the Japanese war ends, we shall have to release our political prisoners. They will find India unsettled and discontented. Food will still be short; demobilisation and the closing down of the war factories, and overgrown clerical establishments, will throw many people out of employment. They will find a fertile field for agitation, unless we have previously diverted their energies into some more profitable channel, i.e., into dealing with the administrative problems of India and into trying to solve the constitutional problem. We cannot move without taking serious risks; but the most serious risk of all is that India after the war will become a running sore which will sap the strength of the British Empire. I think it is still possible to keep India within the Commonwealth, though I do not think it will be easy to do so. If we fail to make any effort now we may hold India down uneasily for some years, but in the end she will pass into chaos and probably into other hands.

P Moon (Ed), pp. 94–5, 98.

Question • Is Wavell enthusiastic about the prospect of Indian independence?

Source 8.15 Arguments for early Indian independence, 1946

5. We are all agreed that we must continue to govern and deal with disturbances up to the point where a general outbreak officially sponsored by Congress begins. The first point for decision is whether we propose to make an attempt to repress a mass movement sponsored and directed by Congress and maintain existing form of Government for a further period . . . Our belief is that reinforcements of British troops on the scale necessary would not be available. None of us feel that the conditions he predicates would be politically acceptable to our supporters and some of us would not personally be prepared to stand for the extensive measures of repression which would be required. Even if they were agreed to the Viceroy considers it doubtful whether they would succeed. The Commander-in-Chief's view is that the Indian Army would disintegrate if it were called upon to deal either with a full scale Congress revolt or a declared Muslim League Jehad. The civilian Services are tired and discouraged and the loyalty of the Police would be uncertain under the strain which a repressive policy would put upon them. Our advice therefore is that the success of a full scale policy of repression is so doubtful that on that ground alone we cannot recommend it. Apart from its feasibility moreover it would mean an end of political progress for a long period and the long term incarceration of the Indian leaders. Government would have to be carried on by Executive Councils of officials both at the Centre and in the Provinces in deteriorating conditions for a long period. We should therefore gain no ultimate advantage. Finally the process of repression might bring as great disasters to India in regard to famine as any other course, and if so we should be held responsible.

8. We have considered but rejected the possibility that we might offer to the Muslim leaders to withdraw into Pakistan and remain there permanently giving to Pakistan economic and military support without which it would not be viable. We are advised by the Commander-in-Chief that on a long term view the military position of Pakistan as a separate sovereign State would be weak and that its defence would cost as much as the defence of the whole of India and would therefore be a long term drain on our resources with little advantage to us. The Viceroy agrees with this view and we reject it on this ground apart from political considerations.

N Mansergh (Ed), pp. 748–51.

Question • On the basis of this document, could it be argued that the British were pushed out of India?

Source 8.16 The French Assembly debates the future of Indo-China, 1947

THE PRIME MINISTER (RAMADIER): Ladies and gentlemen, two things are certain, affirmed by all and should be considered as unanimous conclusions of this debate.

The first is that France must stay in Indochina, that her succession there is not in question and that she must carry out there her civilising work.

(Applause from the left, center and right.)

The second is that the old ways are over . . . Faced with the situation in Indochina we have wanted to have a policy of understanding. . . . Perhaps we have been carried away too quickly by the events; we have treated, we have negotiated. For almost a year, from March 6 until the morning of December 19, we have not ceased to negotiate.

THE OVERSEAS MINISTER (MOUTET): Very good!

THE PRIME MINISTER: And then, on the 19th of December, we were obliged to recognise that the negotiations, even the accords, did not settle everything and that certain acts of violence, of savagery, tore up all the accords and ruined any thought of conciliation.

I have here, ladies and gentlemen, photographs of the mutilated bodies found in the burned-out houses of Hanoi. But perhaps more than any other act of savagery, the total destruction of the Pasteur Institute is the symbol of a wild, destructive will which is not, I am sure, that of the Vietnamese people, but of certain men.

We were then obliged to fight.

Everyone has accepted this struggle, from the time of Leon Blum's Government just as at the moment, on January 21, when I made my declaration [of investiture] before the National Assembly.

(Applause from the left, center and right.)

At that time I said: 'We cannot accept that the peace be disturbed. We must protect the life and the goods of those Frenchmen, foreigners and our Indochinese friends who put their confidence in French liberty. We must assure the security of our forts, reestablish essential communications and insure the security of the people who seek refuge with us.'

This we have done . . .

And it is here that I want to render homage to our Far East Expeditionary Forces, to those soldiers who fight in a climate they are not used to, with a French faith which allows them to hold and to advance under the most difficult circumstances.

They have shown their heroism. France should express her gratitude.

(Prolonged applause from the left, center and right. The deputies seated on these benches rise.)

(Numerous voices from the right and center call to the extreme left [i.e., to the Communists] 'Get up!' From the center: 'It is a disgrace to remain seated!')

RENÉ PLEVEN: I see that the Minister of Defense [Billoux, a Communist] has not risen.

(Exclamations from the extreme left.)

FLORIMOND BONTE [COMMUNIST]: It is we who defend the lives of our soldiers. We are not at your command!

(Exclamations from the right.)

THE PRESIDENT: The Prime Minister has the floor. It is to him alone at this time to express the sentiment of the Assembly and the country.

THE PRIME MINISTER: France should render homage to those who have fallen. France should render homage to those who fight and who continue to fight.

(*Applause from the left, center and right. The deputies seated on these benches rise again.*)

. . . There is no solution but a political solution. Now we must build. And we know what it is we must build. We know because the Constitution says it. . . . Within the framework of the French Union we are going to create and organise an association of nations and peoples who will put or coordinate in common their resources and their efforts in order to develop their respective civilisations, increase their well-being, and insure their security. Not domination, not subjugation: association. We are not masters who have just spoken as masters. We want to be instructors and counsellors. We are associates. This implies that we respect the independence of peoples. The preamble of the Constitution prescribes it: 'Faithful to her traditional mission, France intends to guide the peoples for whom she has responsibility into freedom to administer themselves and conduct their own affairs democratically.'

That is why, last January 21, I stated before you: independence within the framework of the French Union, that is the right to conduct and administer democratically their own affairs, to choose their government, to fix the framework in which the Vietnamese people want to live . . .

But this cannot be realised except within the framework of the French Union which, in addition to such broad freedom, implies mutual obligations as well. The 62nd article of the French Constitution states that 'the members of the French Union will pool all their resources in order to guarantee the defense of the entire Union. The Government of the Republic shall see to the coordination of these resources and to the direction of the policy appropriate to prepare and assure this defense.'

T Smith (Ed), *The End of the European Empire*, pp. 124–7.

Question • What is the significance of the terms 'left', 'centre' and 'right'?

French trenches at Dien Bien Phu, 1954. The soldiers are resting between attacks from the Viet Minh. The French eventually lost their stronghold.

9 *Evaluation of Imperialism in Asia*

A BLESSING OR A CURSE?

"DISPUTED EMPIRE!"

Orient!
The soil on which
naked slaves die of hunger.
The common property of everyone
except those born on it.
The land where hunger itself
perishes with famine!
But the silos are full to the brim,
full of grain—
only for Europe.

Nazim Hikamet, 1925

WAS ASIA BETTER OR WORSE OFF as a result of the colonial experience? This chapter looks at the 'balance sheet' of colonial rule—its achievements, its failures, its limitations. The conclusion is that colonialism generally played an important and perhaps indispensable role as a catalyst of Asian modernisation and cultural renaissance.

KEY DATES

1760 The Industrial Revolution begins in England.
1799 The Dutch East India Company declared bankrupt.
1839 Contraband (illegal) opium imported to China from India reach 40 000 chests a year.
1870 Piped drinking water provided for Manila.
1876 Rubber tree smuggled out of Brazil by English adventurer Henry Wickham and planted in Kew Gardens, London. Seedlings from this tree were used to start the Malayan rubber industry.
1890 Pasteur Institute of Tropical Medicine established at Saigon.
1896 Bubonic plague epidemic in India kills 30 000 people.
1897 (Sir) Ronald Ross (1857–1932) of the Indian Medical Service, working in Calcutta, discovers that malaria is spread by mosquitos.
1898 Execution of Chinese reformer T'an Ssu-t'ung (b. 1865) by order of Dowager Empress Tzu-hsi.
1907 Tata Iron and Steel Company, set up largely with local capital, begins operations at Jamshedpur, India.
1935 First senior high school opened in Cambodia.
1943 Famine and disease kill 2–3 million in Bengal.

ISSUES AND PROBLEMS

Was colonialism in Asia a blessing or a curse? The answer to that question will depend in part on what places and periods are examined. Generally, it would be fair to say that European influence until the 19th century was more destructive than constructive. After that, the record becomes better, particularly in the 20th century. Likewise, the British and especially the Americans did more in the long term for the welfare of the peoples living under their rule than the Dutch, the French, and the Portuguese in particular. However, the answer to the question will be determined mainly by what we think European imperialism in Asia *ought* to have accomplished. Should the Europeans be judged solely by their own standards, and their performance measured against the goals they set themselves? Or should they be judged according to how far they succeeded in 'developing' their

Asian colonies and improving the condition of their inhabitants?

If we take the first approach, allegations of oppression and exploitation become irrelevant. Indeed, the only issue that really need be considered is whether imperialism was 'profitable' to the metropolis. However, that is a less simple task than it may appear at first sight. Calculating the 'drain' to Europe gives only a very incomplete picture. Probably the clearest case of Europe profiting at the expense of Asia was the Cultivation System which the Dutch operated in Java (source 4.12), which is said to have grossed The Netherlands 781 million guilders between 1830 and 1874 (enough to wipe out the national debt and build a modern railway network). However, that spectacular success needs to be weighed against both the heavy losses incurred by the Vereenigde Oost-Indische Compagnie in the 18th century which led to the Company being wound up for bankruptcy, and the large annual deficits that Indo-China cost the French Ministry of Colonies. Moreover the profits (and losses) that accrued to Europe from colonialism cannot be assessed purely in monetary terms. Thousands of lives were lost in acquiring and maintaining European hegemony: surely, a heavy cost. For the French, in particular, the possession of an overseas empire was a vital source of national pride, and the mastery of India, with its standing army of 200 000, was important to Britain's stature and influence as a Great Power. The effort of the missionaries must also be considered. By the start of the 20th century the Christian population of Asia was approximately 14 million. From a missionary point of view, even one soul 'saved' from 'eternal damnation' was doubtless a cause for congratulation. The historian is entitled to ask, however, whether 14 million converts (from a population quickly approaching 1500 million) was really a good return for 300 years of continuous proselytisation.

If we take the second approach, which most scholars would adopt, and evaluate colonialism by what it did for Asia, the picture initially looks even bleaker. Everywhere, Western mass-produced manufactures undercut and gradually destroyed indigenous handicrafts, and China became a gold-mine for drug-traffickers. In the colonies, taxes increased, and the gap between rich and poor grew more pronounced. Yet if colonialism brought exploitation and social dislocation to Asia, it also brought modernisation. By the end of the colonial period most large towns had been linked by railways or metalled roads; agriculture had been transformed by cash-cropping; epidemic disease had been checked by vaccination; custom had given way to law; and science and mathematics had become a fundamental part of the school curriculum. All over the world modernisation has involved hardship, and Asia was no exception. As the British Colonial Secretary, Joseph Chamberlain, bluntly observed in 1897, 'You cannot have omelettes without breaking eggs'.[1] Nevertheless, it is valid to ask whether the painful process of modernisation would have been less traumatic if it had been carried out by native agency rather than by foreigners. Would Asia have developed faster, and more **equitably**, without the European presence?

equitably
fairly, justly

Moreover, scholars who want to prove that the European impact on Asia was completely destructive must first confront the paradox of a mushrooming population. Around 1800 the population of India was about 150 million; a century later, it was in excess of 300 million. During the same period the population of Java and Madura rose from about 3 to 28.4 million, and by 1930 it was over 40 million. Could this phenomenal growth have taken place without, at very least, a massive increase in the supply of food?

THE DEBATE

There are broadly four schools of thought about the Western impact on Asia. The first, made up largely, though not exclusively, of Marxist scholars, thinks that the European presence was almost totally destructive. With the commercialisation of Indian agriculture, writes Anupan Sen,

> a large class of parasitic landowners, moneylenders and land speculators came into existence and more and more people were drawn to these sources of income. Furthermore, since British rule destroyed the urban industries many uprooted people with no other employment fell back on agriculture, which in this way became the only source of livelihood for most of the people of India.
>
> Thus the legendary poverty of India today—in contrast to its legendary riches in the past . . .—is the result of colonial rule.[2]

deindustrialisation
process by which emphasis on cash crops and raw materials replaced importance of manufacturing sector

This may be called the **deindustrialisation** school. The second school, by contrast, believes not only that colonialism was generally benign in its impact on Asia, but that economic development in the region would, to quote Allen and Donnithorne, 'have been slow and hesitating' in the absence of Western leadership.[3] The third school, like the second, is generally supportive of the role which European colonialism played in developing Asia. However, it believes that it should have been bolder in its assault on traditional society. 'If one charge can be levelled against Western imperialism,' concludes I R Sinai, 'it is that it was not revolutionary enough . . . and that it did not, in the essential work of civilisation that it was performing, sufficiently disturb or transform the archaic societies with which it came into contact.'[4] This could be termed the 'aborted modernisation' view. The fourth school goes further. Arguing that significant modernisation under colonialism was confined to the larger port cities, this school concludes that European rule barely disturbed the traditional pattern of Asian life.

Was Imperialism Profitable for the Europeans?

As remarked earlier, calculations of 'profitability' with regard to imperialism and colonialism are extraordinarily complex. It is difficult to place a monetary value on many of the activities which lay at the core of imperialism: for example, the pursuit of power, or the quest for converts. How does one quantify the satisfaction of the European colonial officials at the knowledge that they had ably served their country and, in the process, made life a little easier for thousands of people under their care? Can one really put a price on such things as excitement, blood-lust, awe, and gratitude? The answer must be 'no'. But even if we confine ourselves to things which *can* be measured, such as loot, trade, and industrial production, there remain two problems: first, the lack of reliable figures; and second, the problem of interpreting the data we have gathered.

Consider, for example, the controversial issue of conversion to Christianity. Taken from the authoritative *World Christian Encyclopedia*, source 9.2 gives us an overview of the end result of the missionaries' labours among the Asian people during 350 years. But what do the figures mean? Leaving aside the dubious question of whether all the converts claimed by the Churches were genuine, who is to say what constitutes a successful conversion record? From the missionaries' point of view, every convert to Christianity meant a soul saved. Therefore, no amount of effort was too great, no sacrifice too big, to bring the Gospel to the 'heathen'. Yet the *Times*' Peking correspondent, the **agnostic** Australian George Morrison, points out that the money which was lavished on the missionary enterprise in China and other Asian countries could perhaps have been better spent on other humanitarian works closer to home (source 9.1).

agnostic one who believes that nothing can be proved, one way or the other, about the existence of God

The calculation of economic advantage to Europe arising out of colonialism in Asia is also dependent on interpretation. Marxist writers like Hamza Alavi insist that there was a net 'drain' of wealth from Asia to Europe during the colonial period: the systematic product of loot, 'home' charges (see chapter 4), and profits from export goods funded by land revenue (source 9.3). Indeed, having computed the 'net flow of resources from India to Britain' during the latter 18th century at 2 million pounds a year, Alavi is convinced that the Industrial Revolution was financed by this means. Peter Marshall, however, puts the 'drain' at no more than 18 million for the entire period 1760–1800.[5] Holden Furber thinks that 'it is extremely unlikely that between 1663 and 1793 the proceeds of the sale of home cargoes exceeded the costs of their procurement'. In other words, he claims that the East India Companies operated at a loss! As for the 'home charges', it can be argued that these were a legitimate item of administrative expenditure (see source 4.13). White officials may, in general, have been overpaid, but as individuals they were surely entitled to their pensions.

Again, recent research by David Fieldhouse and others has led to a questioning of the traditional view that colonies were a preferred destination for goods and investments. While the trade of the South-East Asian colonies was mostly with their respective mother-countries, the reverse was not always the case (source 9.4). For Example, Britain bought and sold more from the Americas than it did from its tropical dependencies.

Finally, one must consider the civil and military cost of running the colonies. Although some costs were met locally, others came out of metropolitan budgets. During the first decade of the 20th century the French colonies cost Paris, according to official figures, an average of 105 million francs a year. Evidently, for France, the political and strategic advantages of owning colonies outweighed the financial liabilities.

Source 9.1 A cost-benefit analysis of Christian conversation, 1895

During the time I was in China, I met large numbers of missionaries of all classes, in many cities from Peking to Canton, and they unanimously expressed satisfaction at the progress they are making in China. Expressed succinctly, their harvest may be described as amounting to a fraction more than two Chinamen per missionary per annum. If, however, the paid ordained and unordained native helpers be added to the number of missionaries, you find that the aggregate body converts nine-tenths of a Chinaman per worker per annum; but the missionaries deprecate their work being judge by statistics. There are 1511 Protestant missionaries labouring in the Empire; and, estimating their results from the statistics of previous years as published in the *Chinese Recorder*, we find that they gathered last year (1893) into the fold 3127 Chinese–not all of whom it is feared are genuine Christians—at a cost of £350 000, a sum equal to the combined incomes of the ten chief London hospitals . . .

G E Morrison, pp. 5–6.

Question • Is a 'cost-benefit' analysis applicable to a spiritual enterprise?

Source 9.2 Proportion of Christians in the population of selected countries, 1980

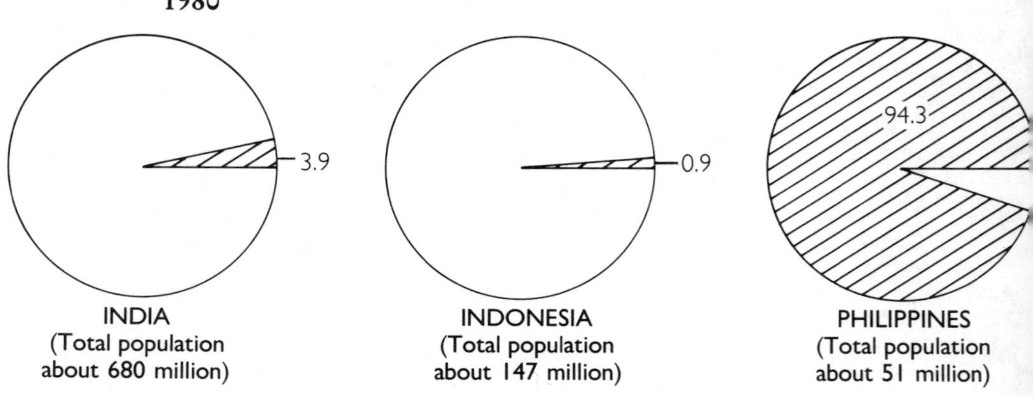

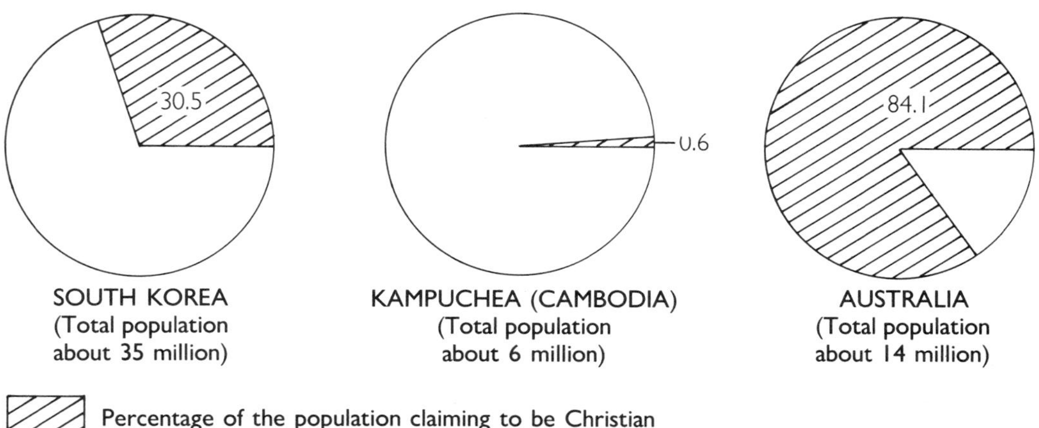

D B Barrett (Ed), pp. 370, 382, 562.

Question • Why was the Christian missionary enterprise so much more successful in the Philippines than in other Asian countries?

Source 9.3 The Indian drain and the English Industrial Revolution

The flow of resources from India to Britain, the 'Economic Drain', was India's 'aid' for Britain's industrialisation. Eric Williams has put forward an argument that profits from the colonial 'triangular trade' financed the industrial revolution in Britain. Britain sold textiles in Africa to finance the capture of slaves who in turn were sold at a great profit in the West Indies. This much-enhanced capital was used to buy sugar which was imported into Britain, multiplying the original investment many times. Williams estimated the profits from this trade to amount to £14 000 in 1739, increasing to £303 000 by 1759.

Economic historians have treated Williams's argument with derision, for the scale of that flow of resources was small in comparison with the level of capital formation during the Industrial Revolution. However such an argument takes on a whole new meaning if we take into account the net flow of resources from India to Britain which I have estimated to be of the order of £2 million annually (Alavi, 1982, 63). That may have been an underestimate for other estimates are higher. Mukherjee quotes Martin writing in 1838: 'For half a century we have gone on draining from two to three and sometimes four million pounds sterling a year from India' (Mukherjee, 1974, 380). Likewise, Prinsep, writing in 1823, estimated the net flow at between £3 million and £4 million annually between 1813 and 1822 (Prinsep, 1971).

Taking even my own lower estimate of an annual flow of £2 million during the critical period of the Industrial Revolution, we find that such a figure is by no means derisory, compared with estimates of capital formation in England during the relevant period, made by Pollard, Crouzet and Deane, which I have discussed elsewhere (Alavi, 1982). Crouzet's estimates, higher than the others, put gross capital formation in the British economy 'in the exceptional boom years of 1790–93' at a total of £9.4 million of which investment in machinery was £2 million, and additional investments in stocks another £2 million (Crouzet, 1972, 33). As against these figures, we put in the balance the annual flow of £2 million (or more) from India, not to speak of large resource flows into Britain from other colonised regions. It is reasonable to conclude that the wealth of Britain, and its industrial might, has been founded on colonialism.

H Alavi & J Harriss (Eds), pp. 11–12.

Question • How did this 'net flow of resources from Britain to India' come about?

Source 9.4 The import–export trade of South-East Asia by origin/destinatio

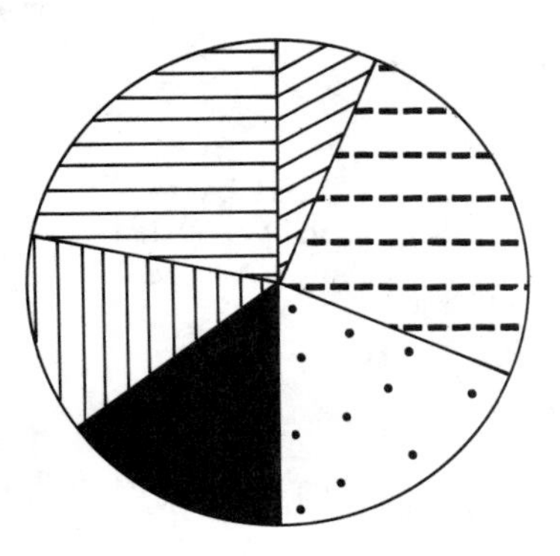
NETHERLANDS EAST INDIES

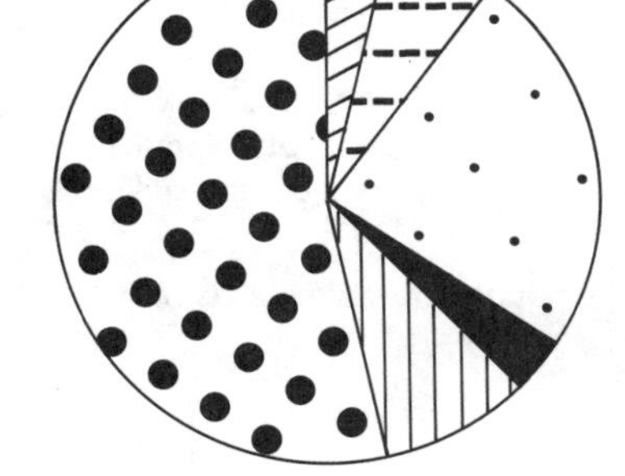
FRENCH INDO-CHINA

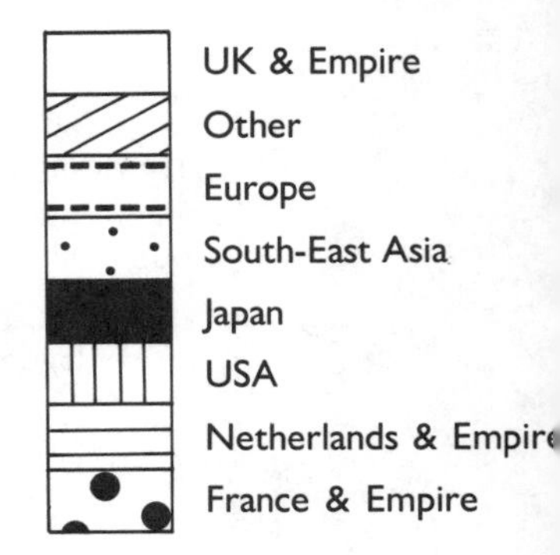

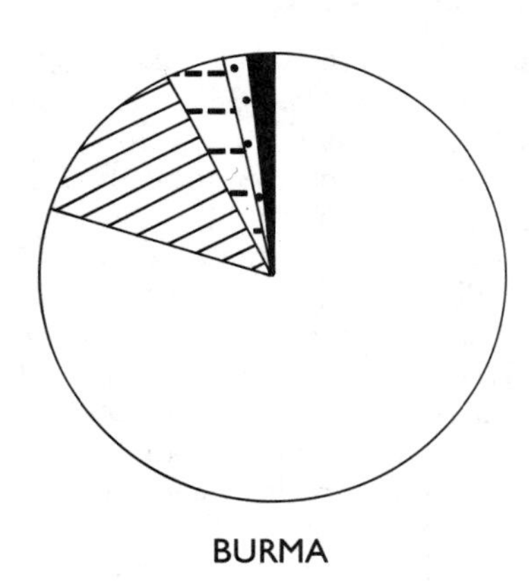
BURMA

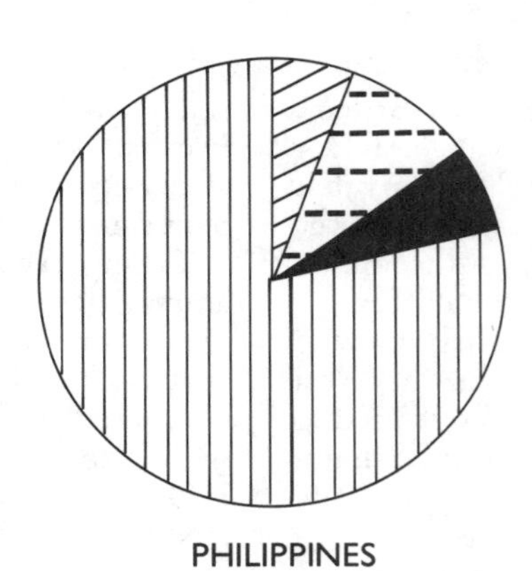
PHILIPPINES

Adapted from D J M Tate, p. 2

Question • Why do you think that The Netherlands' share of Indonesia's trade was proportionately less than Britain's share of Burma, or the USA's of the Philippines?

Did the Natives' Condition Improve?

Assessments of the European impact on Asia also vary greatly, depending on which aspects of colonialism are given prominence. For instance, it was possible for an administrator like Sir Michael O'Dwyer to look back over his 25-year association with the

Punjab with legitimate pride. After all, had not the British brought law and order? Had they not made the deserts bloom through irrigation? The figures which O'Dwyer cites in his memoirs suggest that, at least in this province, British rule had proved something of a blessing (source 9.5).

Was the Punjab typical in this respect? Generally speaking, it was not. Due to extensive canal irrigation, the Punjab's output of wheat increased enormously, but food crop production throughout the rest of the subcontinent struggled to keep up with the growth in population (source 9.6).

Whatever its other achievements in Asia, colonialism certainly did not eliminate poverty. On the most favourable forecast, the colonial period saw a marginal increase in the per capita level of gross national product. But Paul Bairoch and other scholars have shown that the standard of living enjoyed by Third World peoples fell dramatically, after 1750, compared to that of Europeans and North Americans (source 9.7).

Bhurpore Mission women, 1902. Do you think that the condition of these native women was improved by the actions of the British? Which of the four schools of thought regarding the West's impact on Asia —deindustrialisation, benign, aborted modernisation or minimal disruption —does this picture support?

Source 9.5 Sir Michael O'Dwyer on 70 years of progress in British Punjab, 1925

When we took over the Province seventy-five years ago it was agriculturally in much the same condition as Alexander had found it over two thousand years before. Agriculture was most precarious except in the few favoured tracts with a good rainfall or suitable for irrigation from wells; famines of the most devastating nature were of frequent occurrence; nearly all the high-lying lands away from the rivers were arid wastes where the scanty rainfall sufficed only to raise a few patches of crops and luxuriant pasture for a few months in the year; much of the population was still nomadic and predatory. In the lowlands, the canal irrigation was limited to a series of rough cuts which drew off the water in the summer from the rising rivers swollen by the melting snows of the Himalayas, and spread it over the low riverain lands. It did not amount to more than three hundred thousand acres in all.

To-day, as the result of the continuous and combined labours of our engineers and revenue-officers, one after another of the great rivers of the Province have been harnessed to the service of agriculture; great dams have been thrown across them, and the fertilising waters which used to flow uselessly into the Indian Ocean or the Bay of Bengal, are now spread in an increasing flow over the arid uplands, transforming them into expanses of rich cultivation supporting millions of industrious peasants. Today, of the 27 million acres under tillage in the Punjab, there are 11 million acres, or 40 per cent of the total, irrigated from canals—almost entirely constructed by British engineers and financed by British capital—and the crops raised on them in a year are valued at 50 millions sterling. The Punjab irrigation system, built up in the last seventy years, is already twice as great as that of Egypt, which is the product of at least six thousand years; and if and when the present great projects are carried out—and they certainly will not be carried out under a Swaraj [Indian] Government—the irrigated area will be 20 million acres.

Such is the work of a Government which, according to Mr. Gandhi, Mrs. Besant, and the majority of Indian politicians, has drained away the life-blood and the riches of India. I myself saw land in Montgomery and Mooltan selling in 1887 at eightpence an acre. Before I left in 1920 the same land, as the result of canal irrigation and railway extensions, which invariably go hand in hand, was selling at £40 per acre. Can any other country show anything to compare with this wonderful achievement?

M O'Dwyer, pp. 251–2.

Question • How much of this 'green revolution' was due to British colonialism?

Source 9.6 Estimates of average annual per capita output of food and non-food crops in India, from 1893–94 to 1945–46

	Output in index units per capita		*Output of food crops, pounds per capita*
	Food crops	*All crops*	
1893–94 to 1895–96	100	100	587
1896–97 to 1905–06	95	97	560
1906–07 to 1915–16	91	97	547
1916–17 to 1925–26	90	98	538
1926–27 to 1935–36	78	90	461
1936–37 to 1945–46	68	80	399

G Blyn, p. 117

Questions
- What does 'per capita' mean?
- Why is that a more meaningful measure than gross output?
- What are 'index units'?

Source 9.7 Evolution of gross national product (GNP) per capita in the developed world' and Asia, 1750–1980

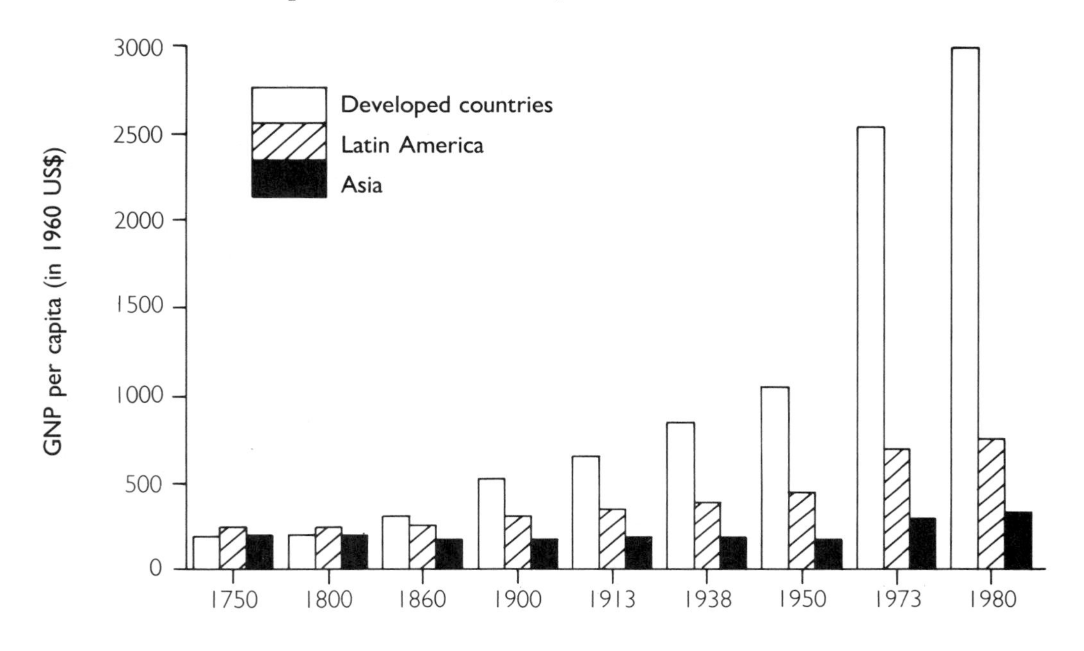

Adapted from P Bairoch, 'The Main Trends in National Income Disparities since the Industrial Revolution', in P Bairoch and M Lévy-Leboyer (Eds), *Disparities in Economic Development since the Industrial Revolution*, pp. 3–17.

Question

- Note that the economic gap between the 'developed' countries and Asia has widened appreciably since 1950, that is *after* the end of formal colonialism. What does this suggest about the link between colonialism and development?

Was Imperialism to Blame for Asian Poverty?

Measured in terms of gross national product per capita, the overall standard of living of the Asian population did not noticeably increase under colonialism. However, it would be hard to deny that for most people the quality of life—with the introduction of the railways, piped water, and modern medicines—did improve. Taxes were higher and there was more bureaucracy, but most Asians in the 19th century would have agreed with the view expressed by the German socialist Karl Marx (1818–83) in an article in the *New York Daily Tribune* in 1853 (source 9.8). Essentially, Marx believed that Western rule was on balance preferable to that which it had replaced.

In any case, it cannot be assumed that the persistence of poverty under colonial rule was a direct result of imperialism, because there are many causes of poverty. Although the actions (or inactions) of governments are important, they are probably less critical than the social factors described by H G Rawlinson (source 9.9). Similarly, while colonial governments helped to smooth the passage for European exports by artificially holding down tariff levels, they cannot be held responsible for the price advantage which Europe and the United States obtained as a result of mechanisation. They should not be blamed either for the historically worsening 'terms of trade' between the industrialised world and the agricultural world, which meant that Asian countries received increasingly lower returns for their **staple** exports (source 9.10). The 'guilty party' here was not European colonialism so much as the rigid law of supply and demand.

staple main item of a country's commerce and trade, often raw materials

Therefore, it is by no means certain that the masses in Asia would have been better off without the colonial yoke. The case of Japan shows that Asian countries were quite capable of modernising without interference from European colonialists. At the same time, the parallel case of China demonstrates that many Asian regimes did not have the desire to change. Although it can be pointed out that people often starved in the colonies, it can be shown that their brothers and sisters in the rest of Asia also starved. Taken from Mallory's 1926 geography of China, source 9.11 shows that famine did not discriminate between nationalities.

Starving natives in a famine-stricken region of India, late 19th century.

Source 9.8 Karl Marx on 'British rule in India'

These small stereotype forms of social organism have been to the greater part dissolved, and are disappearing, not so much through the brutal interference of the British taxgatherer and the British soldier, as to the working of English steam and English free trade. Those family-communities were based on domestic industry, in that peculiar combination of hand-weaving, hand-spinning and hand-tilling agriculture which gave them self-supporting power. English interference having placed the spinner in Lancashire and the weaver in Bengal, or sweeping away both Hindoo spinner and weaver, dissolved these small semi-barbarian, semi-civilised communities, by blowing up their economical basis, and thus produced the greatest, and, to speak the truth, the only *social* revolution ever heard of in Asia.

Now, sickening as it must be to human feeling to witness these myriads of industrious patriarchal and inoffensive social organisations disorganised and dissolved into their units, thrown into a sea of woes, and their individual members losing at the same time their ancient form of civilisation and their hereditary means of subsistence, we must not forget that these idyllic village communities, inoffensive though they may appear, had always been the solid foundation of Oriental despotism, that they restrained the human mind within the smallest possible compass, making it the unresisting tool of superstition, enslaving it beneath traditional rules, depriving it of all grandeur and historical energies. We must not forget the barbarian egotism which, concentrating on some miserable patch of land, had quietly witnessed the ruin of empires, the perpetration of unspeakable cruelties, the massacre of the population of large towns, with no other consideration bestowed upon them than on natural events, itself the helpless prey of any aggressor who deigned to notice it at all. We must not forget that this undignified, stagnatory, and vegetative life, that this passive sort of existence, evoked on the other part, in contradistinction, wild, aimless, unbounded forces of destruction, and rendered murder itself a religious rite in Hindostan. We must not forget that these little communities were contaminated by distinctions of caste and by slavery, that they subjugated man to external circumstances instead of elevating man to be the sovereign of circumstances, that they transformed a self-developing social state into never changing natural destiny, and thus brought about a brutalising worship of nature, exhibiting its degradation in the fact that man, the sovereign of nature, fell down on his knees in adoration of Hanuman, the monkey, and Sabbala, the cow.

S Avineri (Ed), pp. 93–4.

Question • Why does Marx, the enemy of capitalism, apparently welcome the arrival of capitalism in India?

Source 9.9 Sir Henry Rawlinson on the indigenous causes of Indian poverty

Much has been made of Indian poverty, though the estimates of the annual income *per capita* differ so widely that it is difficult to draw any definite conclusions from them. They are invalidated by the fact that a considerable proportion of the population consists of aboriginal tribes, religious mendicants [beggars], gypsies and others, among whom the use of money is almost unknown. In an agricultural country (e.g. England before the industrial revolution) the income per head is necessarily lower than in an industrialised urban population, and in a tropical climate life is simpler and wants are fewer. The tradition of a Golden Age before the advent of the British finds small confirmation from the records of contemporary travellers. It probably owes its origin to the vast hoards accumulated in the treasuries of Indian potentates [rulers], which in the west would have been in circulation. Hoarding, a legacy of the days of the Great Anarchy, is still prevalent among the rural community, despite efforts to popularize agricultural banks, and is one of the causes of impoverishment. The cultivation of cash crops is still a comparatively recent development.

The causes of Indian poverty are to be sought, not in over-taxation and the 'drain,' but in the inherent social structure of Indian society. Caste regulations, with their drastic restrictions, are essentially uneconomic, and, combined with the conservatism of the Indian peasant, make rural prosperity unattainable under present conditions. Attempts at village improvement, like those of A F Brayne in the Gurgaon district of the Panjab, quickly fade away when their originator is transferred to another district. They have no roots in the soil.

Another adverse factor is the phenomenal increase in the population, due to the removal by the Pax Britannica of the natural checks, such as war, famine and disease. Between 1921 and 1931 it was 32 000 000; between 1931 and 1941 it was 51 000 000—an increase of 15 per cent. This means that on an average the rural population is swelled by three millions a year.

H G Rawlinson, pp. 236–7.

Question

- Rawlinson wrote this in 1946. Forty-five years later his excuses for the failure of colonialism to alleviate poverty seem less convincing. Why?

Source 9.10 Terms of trade between 'developed' and 'under-developed' countries, from 1876–80 to 1973

Years	*Unit values (1900 = 100)* *a. Developed*	*b. Under-developed*	*Terms of Trade (a/b)*
1876–80	94	67	140
1896–1900	84	54	155
1913	100	100	100
1928	134	130	103
1937	118	100	118
1953	250	163	154
1962–63	255	143	177
1967–68	270	155	173
1973 (est.)	385	186	205

M B Brown, p. 25

Question

- Why have the 'terms of trade' moved increasingly in favour of the 'developed' countries since World War I?

Source 9.11 Famine in north China, 1877

In November, 1877, the aspect of affairs was simply terrible. The autumn crops over the whole of Shansi and the greater part of Chihli, Honan, and Shensi had failed. . . . Tientsin was inundated with supplies from every available port. The Bund was piled mountain high with grain, the Government storehouses were full, all the boats were impressed for the conveyance of supplies toward Shansi and the Hochien districts of Chihli. Carts and wagons were all taken up and the cumbersome machinery of the Chinese Government was strained to the utmost to meet the enormous peril which stared it in the face. During the winter and spring of 1877–78, the most frightful disorder reigned supreme along the route to Shansi. Hwailu-hsien, the starting point, was filled with officials and traders all intent on getting their convoys over the pass. Fugitives, beggars, and thieves absolutely swarmed. The officials were powerless to create any sort of order among the mountains. The track was completely worn out, and until a new one was made a dead block ensued. Camels, oxen, mules, and donkeys were hurried along in the wildest confusion, and so many perished or were killed by the desperate people in the hills, for the sake of their flesh, that the transit could only be carried on by the banded vigilance of the interested owners of grain, assisted by the train bands, or militia, which had been hastily got together, some of whom were armed with breech-loaders. . . . Night traveling was out of the question. The way was marked by the carcasses or skeletons of men and beasts, and the wolves, dogs, and foxes soon put an end to the sufferings of any wretch who lay down to recover from or die of his sickness in those terrible defiles. . . . Broken carts, scattered grain bags, dying men and animals so frequently stopped the way that it was often necessary to prevent for days together the entry of convoys on the one side, in order to let the trains from the other come over. No idea of employing the starving people to make a new, or improving the old road ever presented itself to the authorities; and passengers, thankful for their escape from the dangers of the journey, were lost in wonder that the enormous traffic was possible.

J A Harrison, pp. 55–6.

Question

- In 1877 famine also raged in south India. Was the response of the Raj any better.

amines in the Indian
bcontinent, 1837–1944.

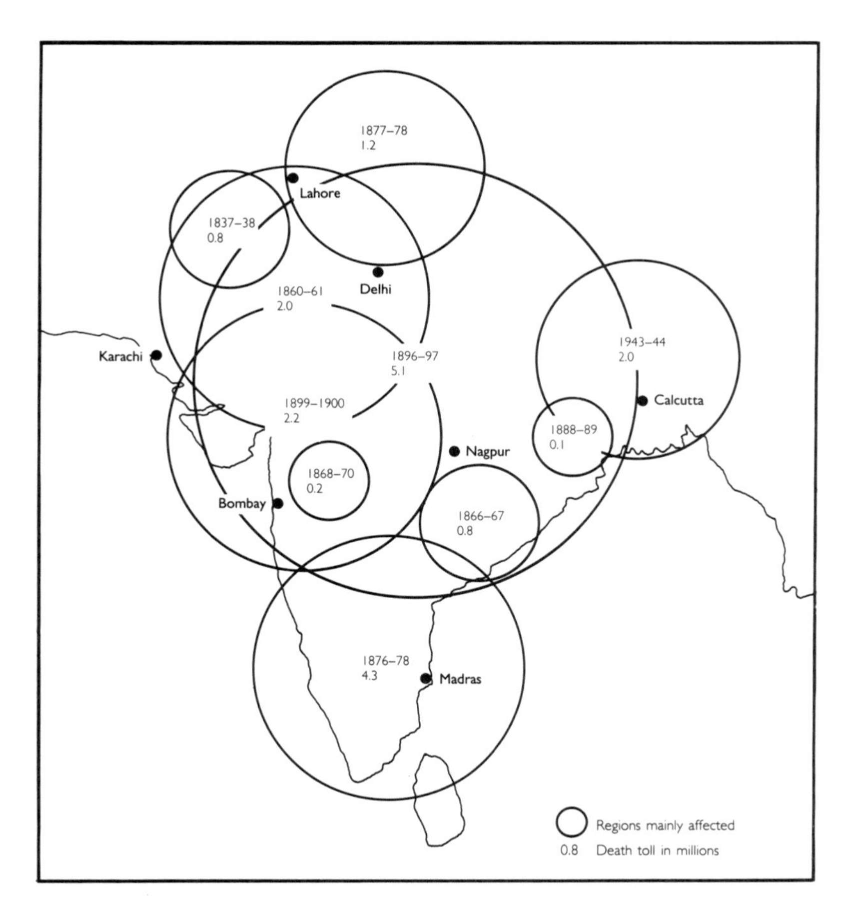

Were the Colonies Underdeveloped?

underdeveloped
term describing 'backward' economic condition of the Third World compared with the First (industrialised) World

Still, there does not seem to be much doubt that the colonial regimes could have done a lot more towards improving the economies of their **underdeveloped** Asian territories if they had been so inclined. Under the British Raj, governmental expenditure was heavily slanted towards furthering imperial ends: most of the money supported the army, police and civil service and to meet interest payments on foreign loans. Comparatively little was spent on things of direct benefit to the people such as schools, roads, hospitals and agricultural research (source 9.12).

However, the British record on development was better than that of the French. The American historian Martin F Hertz points out that Cambodia (a French protectorate, not a fully fledged colony) did not have a single senior high school until 1935 (source 9.13). The country's only tertiary graduates—three teachers, a doctor and an engineer—had all studied in Hanoi or Paris.

Likewise, none of the colonial regimes made a substantial effort to promote industrialisation, for example, by setting up model factories and imposing protective tariffs. Indeed, many scholars would argue that the Western Powers deliberately stifled industrial growth in Asia in order to protect their export markets and to ensure that Asia remained a supplier of cheap raw materials. For instance, Marxists Caldwell and Utrecht conclude that 'Indonesia in particular, and [the nations of] South East Asia in general, have been thwarted from turning their resources into wealth for the peoples of the region by . . . colonialism and neo-colonialism' (source 9.14).

conspiratorial
involved in plotting secretly, often unlawful

However, the *laissez-faire* stand taken by the colonial regimes was not necessarily **conspiratorial**. Coming as they did from a private enterprise system, it simply did not occur to the Europeans that they had any such obligations. Is it fair to blame them for acting consistently with their beliefs?

Source 9.12 Government expenditure in India from 1870–72 to 1946–47

Year	*Total govt. exp. (mill rupees*	*Expenditure on imperial purposes (%)*			*Social and developmental expenditure (%)*		
		Administration	*Debt-service*	*Defence*	*Education*	*Public health*	*Other*
1871–72	467	NA	13	32	NA	NA	NA
1900–01	958	24	4	22	2	2	33
1913–14	1199	27	2	25	4	2	33
1917–18	1335	27	8	33	4	2	17
1921–22	2135	24	8	33	4	2	18
1931–32	1906	28	12	28	7	3	12
1946–47	7973	15	6	26	3	2	33

Cambridge Economic History of India, Vol. 2, p. 93

Question • Could these figures be used to make out a case that the British Raj gradually became more efficient?

Source 9.13 The educational record of the French in Cambodia

The greatest French sin of omission concerned the field of education. An educated Cambodian elite might have aspired to high posts in the country's administration, but the kind of education that was provided stopped well short of preparation for university study. A senior high school was provided only in 1935, and the number of graduates from that institution ('bacheliers') in 1939 was exactly four. There was, of course, no institution of higher learning in Cambodia, and Cambodians were discouraged from attending such institutions in France. Only one Cambodian obtained a medical doctor's degree in France before the war, and he was able to do so because he remained in France after enlisting in its army during World War I. Cambodia's only pre-war graduate engineer, Sonn Voeunsai (at present in charge of the national railways), is the son of that doctor who, having resided in France, was able to overcome the obstacles in sending him there. Not a single Cambodian could study architecture, nor were any trained to qualify for leading positions in the various government departments such as agriculture, the postal service, public works, etc. By an accident Cambodia had one man (Sonn Sann) whose family had been able to have him study at the École des Hautes Études Commerciales in Paris and who thus had some qualifications to head the new National Bank when it was created in 1954. The death of trained executive talent is apalling.

J Bastin (Ed), pp. 104–5.

Question • Is it fair to judge French colonialism just by what it did, or did not do, in Cambodia?

e distribution of
hools in French
lo-China, *c* 1900.

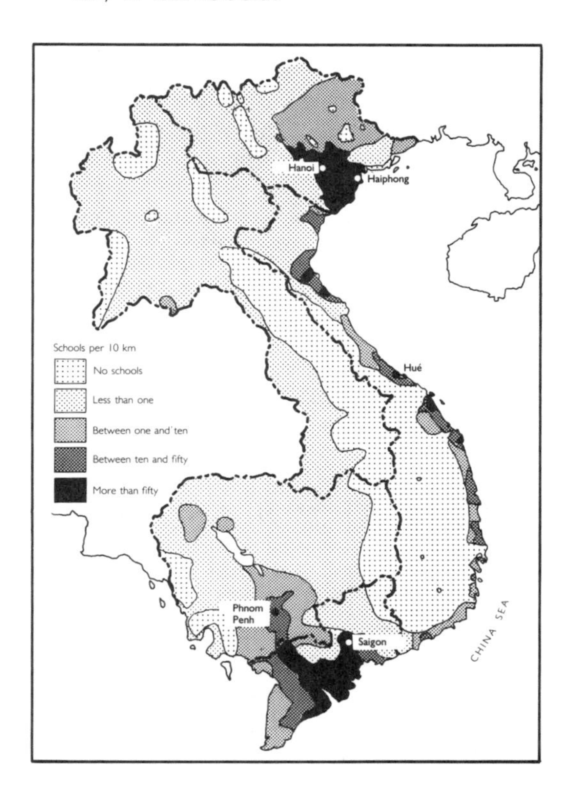

Source 9.14 The supression of native enterprise in Indonesia

With one foot pressing down on the neck of the Javanese sugar cultivator and the other similarly weighing on the Sumatran rubber smallholder, the Hollander had kept aloft a profitable plantation economy: weakened by the depression and felled by the war, he collapsed, and with him the plantation system. It is important to note, however, that had the Javanese sugar grower and Sumatran rubber grower ever had the chance of autonomous accumulation of capital, they might very well—indeed almost certainly would—have founded a sound export sector on a totally different and more secure basis. Admittedly there are all kinds of imponderables here, involving the nature of an independent Indonesian government (socialist or capitalist), world market conditions, the readiness of, say, the Soviet Union to enter into long-term buying or barter agreements, and so on. A priori [reasoning from causes to effects], however, we would be quite wrong to dismiss the hypothesis that, sustained by an export sector in indigenous control, an independent, as opposed to a colonial, Indonesia would have trodden the path pioneered by Japan with silk.

But Dutch power was not only deliberately and directly applied to discourage and hamper Indonesian enterprise. It had in addition many other and more subtle distorting influences on indigenous economic performance and achievement. For instance, one might cite the Dutch-fostered encroachment of the Chinese on so many areas of Indonesian economic life, virtually in the end pre-empting some sectors. Generally speaking, wherever there was apparent a lucrative opening, whether in the salaried or commercial ranks, the 'native' was excluded one way or another. Thus it was that the odds were heavily stacked against Indonesians succeeding in accumulating enough capital and contacts and experience to make the transition from petty trading and handicrafts to creation of an autonomous modern manufacturing sector. That there were investible funds in indigenous hands may, however, be seen from the fact that in 1927, to take but one year at random, 60 000 Indonesians made the pilgrimage to Mecca at a cost per person of about 1000 florins—a total expenditure of the by no means negligible sum of 60 million florins (to put this in perspective, it may be noted that the average income of a Javanese family in 1928 has been put at around 200 guilders). Saving was, therefore, obviously both possible and actually undertaken, a further indication that it was institutional restraints which inhibited achievement of autonomous national economic development in 19th and 20th century Indonesia . . .

Throughout the long colonial period, Indonesia's potential investible surplus was systematically siphoned out of the country largely for the benefit of non-Indonesians . . . The point worth stressing is that Indonesia in particular, and South East Asia in general, have been thwarted from turning their resources into wealth for the peoples of the region by a history of colonialism and neo-colonialism.

M Caldwell and E Utrecht, pp. 46–7.

Question • The authors call their book 'an alternative history'. What is 'alternative' about this extract?

Were the Reasons for Development Distorted?

Another criticism sometimes made of European colonialism in Asia is that it fostered an unhealthy respect for products and ideas made in the West. The result of this was that indigenous flair and creativity was suppressed or distorted. Sources 9.15 and 9.16 express this point of view.

In source 9.15, American traveller Mary H Fee argues that the desire for Western education in the Philippines did not grow out of any real needs that the Filipinos had. In her view, they were perfectly content with their traditional lifestyle. But they were eager to prove that they could compete successfully in the modern world.

In source 9.16, the 20th century Japanese writer Junchiro Tanizaki uses the example of the fountain-pen—one of the many pieces of Western technology that accompanied imperialism around the world—to demonstrate that even a commonplace gadget can have far-reaching cultural implications. Of course, when Junchiro wrote, he had no idea that, in less than half a century, his countrymen would be selling Japanese technology to the West.

Source 9.15 Enforced 'progress' in the Philippines, 1910

A great deal has been said in the American press about the eagerness for education here. The desire for education, however, does not come from any real dissatisfaction which the Filipinos have with themselves, but from eagerness to confute the reproach which has been heaped upon them of being unprogressive and uneducated. It is an abnormal condition, the esult of association of a people naturally proud and sensitive with a people proud and arrogant.

At present the desire for progress in things educational and even in things material is ineffective because it is fed from race sensitiveness rather than from genuine discontent with the existing order of things. The educated classes of Filipinos are not at all dissatisfied with the kind and quality of education which they possess; agriculturists are not dissatisfied with their agricultural implements. If you talk to a Filipino about the carefully constructed houses of America, he does not sigh. He merely says, 'That is very good for America, but here different custom.' Filipino cooks are not dissatisfied with the terrible *fugons* which fill their eyes with smoke and blacken the cooking utensils, and have to be fanned and puffed at every few minutes and occasionally set the house on fire. The natural causes of growth are not widely existent. Meanwhile growth goes on stimulated by eternal critism, the sting of which the Filipinos would move heaven and earth to escape . . .

All the natural laws of development are turned around in the Philippines . . . The Filipinos began the march of progress at a time when the telegraph and the cable and books and newspapers and globetrotters submitted their early development to a harrowing comparison and observation. The Filipino is like an orphan baby, not allowed to have his cramps and colic and to cut his teeth in the decent retirement of the parental nursery, but dragged out instead into distressing publicity, told that his wails are louder, his digestive habits more uncertain, his milk teeth more unsatisfactory, than the wails or the digestive habits or the milk teeth of any other baby that ever went through the developing process.

M N Francisco and F M C Arriola, p. 88.

Question • Assuming the author was right in her observations, would things have been any different if the Philippines had not come under American rule?

Source 9.16 Junchiro Tanizaki on the tyranny of the fountain-pen

. . . had [the fountain-pen] been invented by the ancient Chinese or Japanese, it would surely have had a tufted end like our writing brush. The ink would not have this bluish colour, but rather black, something like . . . India ink, and it would have been made to seep down from the handle into the brush. And since we would then have found it inconvenient to write on Western paper, something near [to] Japanese paper—even under mass production, if you will—would have been most in demand. Foreign ink and pen would not be as popular as they are; the talk of discarding our system of writing for Roman letters would be less noisy . . . But more than that: our thought and our literature might not be imitating the West as they are, they might have pushed forward into new regions quite on their own. An insignificant little piece of writing equipment, when one thinks of it, has had a vast, almost boundless, influence on our culture. . . . [Having] come this far we cannot turn back. . . . [But] it is not impossible that we would have one day discovered our substitute for the trolley, the radio, the airplane of today. They would have been no borrowed gadgets, they would have been the tools of our own culture suited to us.

P Kontos *et al* (Eds), pp. 149–51.

Question

- Can you think of any other everyday object which might look different if it had been invented by the Japanese?

'Confusion of Ideas.' In this lithograph, Rabindranath Tagore, the famous Bengali poet and artist, deftly satirises the one-sided modernisation which occurred in many parts of Asia under the impact of European colonialism. The two Indians show by their clothes that they think of themselves as 'modern, rational men', yet they remain wedded to superstitious rituals.

Did Imperialism Make a Difference?

Writing in the 1840s, the first White Rajah of Sarawak, James Brooke, could find little good to say about European rule in Asia. In India and in the Malay Archipelago, he wrote in his journal, rather than improving the condition of native society, the white man had gone a long way to destroying it (source 9.17). Does this verdict also apply to the later 19th and 20th centuries? Overall, the answer is probably 'no'.

Although the Europeans committed many terrible crimes in Asia during this period, as shown in Chapter 6, their power was not sufficient to enable them to sweep away at will entire social systems, even in those parts of South and South-East Asia that they ruled directly. Certainly, some places and trades suffered extensively from Western intrusion and competition. The port cities and the handicraft sector are examples.

But in the villages life was little changed, despite the introduction of cash-cropping. Traditionally suspicious of outsiders, peasants did not take kindly to bureaucrats from the cities telling them how to run their lives. Furthermore as Thai anthropologist Phya Rajadhon points out in source 9.18, they could see little advantage in adopting new expensive technology when the old methods worked perfectly well.

Those changes which did occur happened basically because of a conscious decision by Asians to seek Western education and to work in close collaboration with Europeans. As shown in an earlier section, the colonial regimes did little to promote industrialisation or public welfare, but this didn't stop Asians from moving in to fill the gap. In India, for instance, groups like the Parsis of Bombay quickly grasped the value of modern machinery and the factory system. By the 1870s a flourishing Indian-owned textile industry was competing with that of Leeds and Manchester. Similarly, in Malaya, the tin mining industry and much of the retail trade was controlled by Chinese. They, not the Europeans, were the real **entrepreneurs** (source 9.19).

entrepreneurs
those who undertake business enterprises

Imperialism and colonialism did make a difference to Asia, but with hindsight one can say that their impact—both positive and negative—was much less substantial than most contemporary observers imagined.

Source 9.17 James Brooke on the effects of European domination in the Malay Archipelago, 1840

Our boasted territory in India, the best and most uprightly governed of any European possession, can, after all, claim but negative advantages. It is neither oppressive nor unjust, and the people are moderately happy; but what advance have they made during the long period of our sway? Are they more civilised than in the time of Baber and of Akbar? Are their minds more enlightened? their political freedom more advanced? their religion less dominant or less bigoted? No: though the English government has used the *best means* to shake the dominion of priestcraft, it still continues. The mass are certainly as ignorant as ever; ignorant of their own rights, content under every or any government, so that they reap the fruit of labour; and, in this respect, are as low as the African!

Lastly, I must mention the effect of European domination in the Archipelago. The first voyagers from the West found the natives rich and powerful, with strong established governments, and a thriving trade with all parts of the world. The rapacious European has reduced them to their present condition. Their governments have been broken up; the old states decomposed by treachery, by bribery, and intrigue; their possessions wrested from them under flimsy pretences; their trade restricted, their vices encouraged, their virtues repressed, and their energies paralysed or rendered desperate, till there is every reason to fear the gradual extinction of the Malay races.

This is the historical record of the rule of Europeans from their earliest landing to the present moment. The same spirit which combines the atrocity of the Spaniard with the meanness of the Jew peddlar, has actuated them throughout, receiving only such modifications as time or necessity has compelled them to adopt. Who that compares the states of the Peninsula, Java, Sumatra, Borneo, or Celebes, before and subsequent to the period of European domination, but must decide on the superiority of the former?

Let these considerations, fairly reflected on and enlarged, be presented to the candid and liberal mind; and I think that, however strong the present prepossessions, they will shake the belief in the advantages to be gained by European ascendency as it has heretofore been conducted, and will convince the most sceptical of the miseries immediately and prospectively flowing from European rule, as generally constituted.

R Mundy, pp. 79–81.

Question • Holding these opinions, why did Brooke allow himself to be made Governor of Sarawak?

Source 9.18 Phya Anuman Rajadhon on the Thai peasant's attitude to change

It is sometimes said that if we changed our plows and used iron plows like those of western countries, we would get better and quicker results than with our old [wooden] plows which have been used for a long time and have never been altered or improved in any way. The explanation [that] is given [for not changing] [is] that western plows are very heavy; oxen and buffaloes can scarcely draw them, and they dig too deep into the earth. Also, they are more expensive than local plows; the farmers usually do not have enough money to buy them. . . . This is a fact, but there is also another fact, namely, that in the case of anything that has been done in the past and has produced visible results, and has long since become custom, not normally manifesting any defects or disadvantages, it is usual among farmers not to want to change, for they do not trust new things, being uncertain that if they change they will receive the expected benefits. If their expectations are not fulfilled, they are in trouble, and so they prefer to do as they have done in the past rather than venture to change to unaccustomed things, unless someone first acts to serve as an example proving that good results [can be] . . . obtained.

P A Rajadhon, p. 309.

Question • In what ways did peasants living under colonial rule respond positively to the challenge of modernisation?

Source 9.19 Dr Lim Boon Keng on the contribution of the Chinese to progress in Malaya, 1917

The Chinese have been the greatest pioneers of civilisation in Malaya. To their perseverance and skill, as well as to their enterprise, is due the extensive development of the peninsula. European trade relies upon their industry and integrity for its wide distribution to the inmost parts of the jungle and for the ultimate barter with barbarian tribes. The Chinese traders in the big towns are the middlemen who conduct an extensive business with the hinterland, carrying the products of civilisation to exchange with the products of nature or with the articles resulting from native local industry. When people feel jealous at the wealth and prosperity of the Chinese communities, they seldom think of the perils which annually destroy thousands of robust men in the prime of life. The interior trade of the Malay Peninsula is attended with multitudinous risks, and holocausts of young lives at the altar of commerce are part of the price paid so willingly by the Chinese for the success of their various undertakings. . . .

L K Wong, p. 67.

Question • Can you find, elsewhere in this book, any documentary evidence which substantiates Dr Lim's claim?

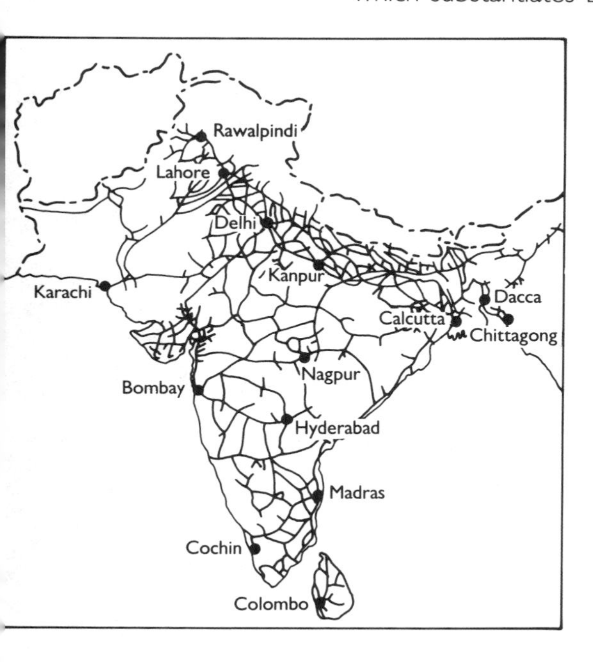

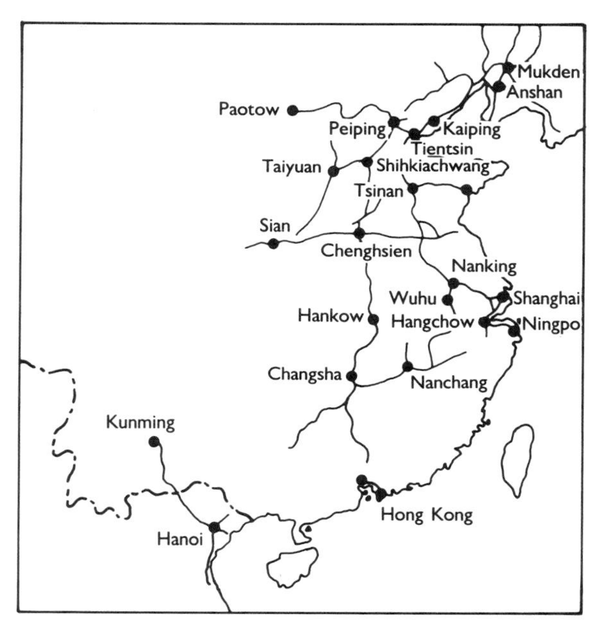

ailways in India and :hina, 1936. India, a olony, is clearly far iore 'developed' than :hina, a semi-colony.

Some Revealing Comparisons

This chapter has argued that colonialism in Asia was notable more for its sins of omission than for its sins of commission. While it brought relative peace, and the rule of law, and created plentiful opportunities for astute Asians to make their mark, it did little by way of economic development and public welfare to reduce the poverty which afflicted the vast majority of the population.

However, this judgement would not be equally true for all periods and for all colonial regimes. As previously remarked, the European record in Asia prior to the 19th century was uniformly bad, but after that it becomes much better. Likewise, the United States and Britain probably did more for the welfare of their subjects than France or The Netherlands. This is certainly the view which American political scientist Frank Darling takes in his book *The Westernization of Asia* (source 9.20).

To some extent this view can be substantiated statistically. Source 9.21 tabulates the extent of road and railway construction in South-East Asia in the 1930s. It will be seen that the British, who ruled in Malaya, Burma and North Borneo, did best in this area; the French, who controlled Indo-China, second best; and the Dutch who held sway in Indonesia, third best. Note, however, that last place goes to Thailand, which was not under colonial rule at all. This supports our earlier contention that colonialism may not have been the main obstacle to Asian development.

Finally, source 9.22 attempts to fit the issue of European imperialism in Asia into a world-wide context. In *The Pattern of Imperialism*, Tony Smith suggests that, until well into the 20th century, Asia remained the single most important arena of British imperial activity. Once Asia obtained its independence, colonialism as a global institution was doomed, and world-wide repercussions were inevitable.

Source 9.20 High-low continuum depicting the degree of transformation induced by the social policies of the colonial Powers

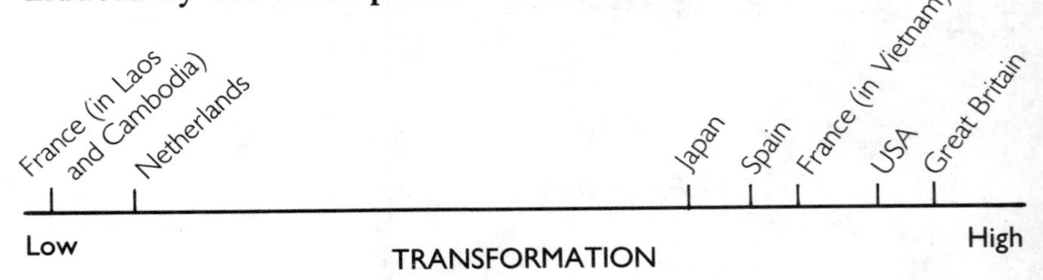

F Darling, p. 140.

Question

- On the basis of your reading, do you agree or disagree with Darling's assessment?

Source 9.21 Roads and railways in South-East Asia in the 1930s

Country	*Population (millions)*	*Railways (k'metres)*	*Roads (k'metres)*	*Total (k'metres)*	*K'metres per head*
Malaya & N. Borneo	5	1894	10 373	12 267	0.0025
Burma	16.8	3296	27 304	30 600	0.0018
Indo-China	23	2098	35 656	37 754	0.0016
N. East Indies	62	7339	51 000	58 339	0.0009
Philippines	15.9	1352	24 477	25 799	0.0016
Thailand	14.5	3418	3 587	7 005	0.0005

D J M Tate, p. 27.

Questions

- Which colonial power comes out best in this survey?
- Why?

Source 9.22 British commerce by origin/destination as a percentage of total British trade, 1860 and 1913

	1860		*1913*	
	Imports	*Exports*	*Imports*	*Exports*
Latin America	9	9	10	11
Africa (total)	8	4	6	10
South Africa	1	1	2	4
Asia (total)	14	17	12	24
India	8	11	6	13

T Smith, *The Pattern of Imperialism*, p. 27.

Question

- Do these figures support the Hobson–Lenin theory that late 19th century expansion was motivated by economic ends?

Epilogue

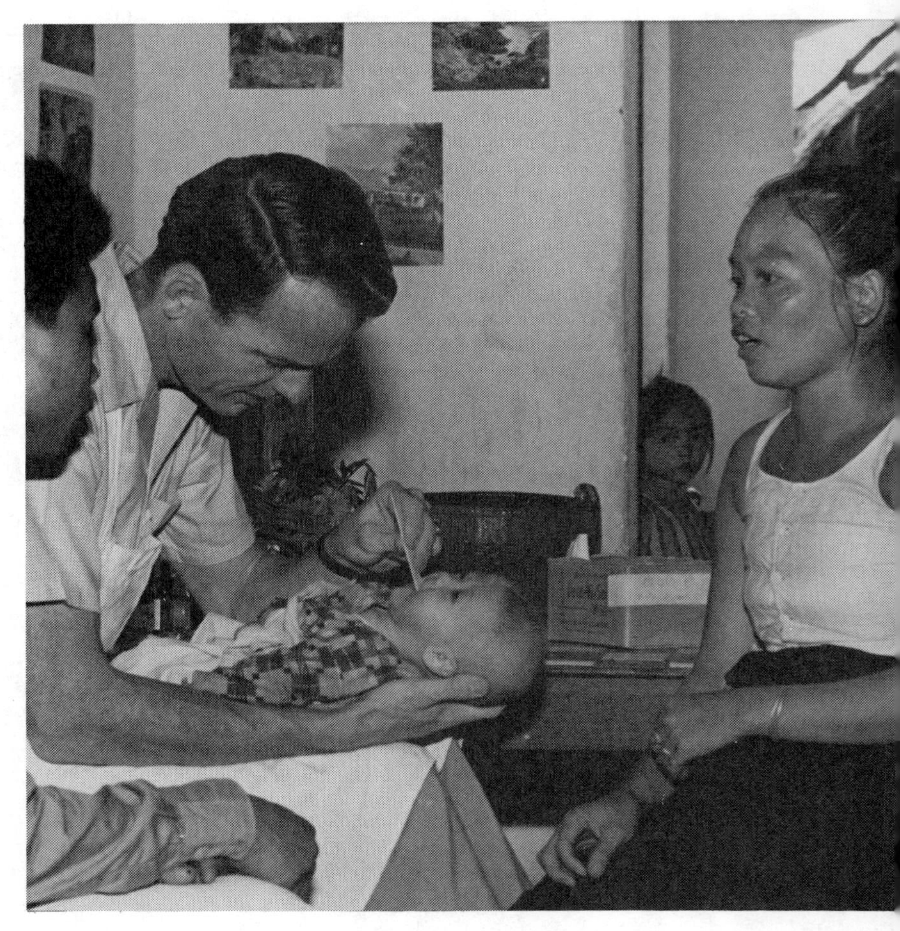

How ridiculous it is, this sickly world! The missionaries said, 'put on these clothes'. Now its the Europeans who come back as tourists [who are] half-naked, topless.

Sir Kamisese Mara, 198

The Colonial Legacy

The half century or so that has elapsed since the end of World War II has witnessed the dismantling of the European colonial empires in Asia. Between 1945 and 1960 more than 40 countries, with a combined population of 800 million, became independent. The few tiny imperial territories which still remain—scattered remnants of a once mighty institution—are in the process of being handed over to their rightful owners. Brunei, in Borneo, was returned to its dynastic ruler, the Sultan, in 1984; Hong Hong is scheduled to revert to Chinese control in 1997; and Macao, the 15.5 square kilometre **enclave** across the water, occupied by Portugal since 1557, is due to be relinquished on 20 December 1999. Perhaps fittingly, the first Europeans to arrive in this part of the world, the Portuguese, will be the last to leave.

enclave foreign territory surrounded by one's own territory

The passage of time has also removed from the scene most of the surviving political figures of the colonial era. Jawaharlal Nehru, Gandhi's lieutenant in the struggle for freedom against British rule and India's first Prime Minister, died in 1964. Ho Chi Minh, the architect of Vietnam's freedom, died in 1969. Sukarno of Indonesia died in 1970. China's 'Great Helmsman', Mao Zedong, died in 1976. Lord Mountbatten, the last Viceroy of India, was assassinated (ironically, by Irish nationalists) in 1979. Sir Malcolm Macdonald, who presided over the transfer of power in Malaya, died in 1981. And the Showa Emperor Hirohito's 66-year reign (the longest in Japanese history) came to an end in 1989. In the former colonies, the statues of European monarchs, statesmen and soldiers which once graced or dominated (depending on your view) city squares and gardens have virtually all been demolished. The street names, which once spoke of London and Paris, now celebrate local myths and heroes.

Ho Chi Minh at 67.

Gateway to the President's Palace, Parliament Square, New Delhi.

Superficially, the European era in Asia would seem to be over, dead and (some would say deservedly) buried. But that is not the case. Although the colonies are gone, much of the imperial administrative framework survives in one form or another. The great railway and canal networks still exist, as do many colonial public buildings. The Viceroy's residence in New Delhi, built to a design by Sir Edward Lutyens, now houses the President of the Republic, and the century-old Raffles Hotel remains Singapore's most familiar landmark. In addition, most of the former colonial boundaries have survived, although they are often at variance with ethnic and cultural patterns. The western part of New Guinea, Irian Jaya, is culturally part of Melanesia. Being a former Dutch territory, however, it is within Indonesian boundaries. Likewise, India has consistently rejected China's claim to Himalayan territory stolen by the British Raj from Tibet in 1910. Internal political contests, too, sometimes bear witness to a colonial past. The struggle of the Karens, Mons and Shans to secede from Burma goes back to World War II when these tribal groups were recruited by the Indian Army to fight against the Japanese and their Burmese collaborators. The rivalry between the two main political parties in Sabah—the Christian Parti Bersatu, and the Malay-Muslim party of the long-serving Chief Minister Tun Mustapha Harun—has its roots in the indigenous Kadazans' conversion to Catholicism during the latter 19th century. Last but not least, there has been a substantial constitutional carry-over. In the large majority of Asian countries the Western system of parliaments, elections, cabinets and courts remains in force. In some countries (for example, India and Pakistan), there has been a conscious effort to incorporate indigenous elements (in the one case, the traditional system of village *panchayats*, in the other the Islamic legal code). But even these countries have not, to any significant extent, abandoned the imported system. Indeed, more than 250 articles of the Indian Constitution of 1950 were lifted straight from the Government of India Act of 1935 that was imposed by the British.

The Global Village

The colonial legacy extends far beyond politics, however. One of the most poignant reminders of the long European presence are the so-called 'Eurasians'—people of mixed blood—who are to be found in their thousands in the towns and cities of south and South-East Asia. In the small Japanese town of Kure, for example, there are no less than 85 children who proudly bear English names—sons and daughters of 'war brides' of Australian members of the Allied force which occupied Japan from 1945 to 1952. Asian Christians are another living legacy of imperialism and they are now estimated to number at least 200 million. Like their parent countries, the Asian Churches now have autonomy in the management of their internal affairs.

But clearly the most important cultural artefact that Asia has inherited from the colonial era is language. English, French and Dutch were introduced initially to serve the needs of the colonial administrations. They were then taken up enthusiastically by Asian élites, who saw in them the key, not only to government employment, but also to Western philosophy and science. The new knowledge that was systematically and irrevocably reshaping the physical and mental world was made available to them. Indeed, by the later part of the colonial era, it was no longer considered enough for an aspiring professional person or politician in Asia to study Western learning at home. For those who could afford it, the highest achievement became an education at a leading European or American university. Jawaharlal Nehru, Singapore's Lee Kuan Yew and Tengku Abdul Rahman, the first Prime Minister of Malaya, all studied at Cambridge. Sir S W Bandranaike of Ceylon went to Oxford. Indonesia's Sutan Sjahrir took out a medical degree at Leiden. T V Soong, Finance Minister under Chiang Kai-shek, graduated from Harvard. The popularity of English, at least, has not measurably diminished since the departure of the colonial regimes. On the contrary, with the expansion of educational opportunities generally, the total number of people literate in one or more Western languages has actually increased since 1950. In India alone, the daily circulation of English-language newspapers is now in excess of 3 million. Moreover, as well as surviving vigorously in their own right, English and other European languages have left their mark on the vernacular languages of the region. Hindi and other Indian languages have incorporated a number of English words; Indonesian and Malay have borrowed extensively from Dutch and Portuguese; and Vietnamese, Tagalog, Turkish, Indonesian and Malay have all adopted the Roman script.

Vestern cultural images
Bhopal, India.

The global acceptance of Western science and technology does not really need much explanation. The fact that a diesel locomotive is made in Birmingham or a silicon chip manufactured in California does not alter their usefulness to the consumer: to the extent that it can offer mankind a better livelihood, the appeal of modernisation is truly universal. However, the Westernisation of the non-European world has not stopped at the limits of invention. As we have seen, Western ideas have also taken root, and so, more surprisingly, have many aspects of Western popular culture, such as food, dress, music and sport. 'Fast food', including beef-burgers, has become an integral part of the Asian diet; Coca-cola is available in China; teenagers in India are willing to pay high prices for second-hand pairs of Levis; and in Japan, homes that once had few internal partitions (those that they did have were paper-thin) are now commonly equipped with solid walls and doors with locks. In the lead-up to Ramadan, the month-long Islamic festival during which no food or drink can be consumed between sunrise and sunset, planes from Pakistan are filled to capacity with wealthy businessmen eager to escape to less puritanical lands in the West. Ask a Sri Lankan what sport he likes, and he will almost certainly say 'cricket'. Ask a Japanese, and he will probably say 'baseball' or 'golf'. Truly the world has become, in Marshall Macluhan's eloquent phrase, a 'global village'. But it is not, yet, a village of equals. In cultural terms, the West still dominates. As the French historian Herbert Luthy observed some years ago,

> Colonisation . . . has come to an end, but the Europeanisation of the world, superficial and distorted as it may sometimes appear, is progressing more rapidly than ever.[1]

'Travolta Sahib' in Bombay, India.

Neocolonialism

> **neocolonialism** term describing imperialist actions of Western agencies in Asia after World War II

As well as flooding Asia (and the rest of the world) with popular culture, the West continues to exert considerable direct and indirect influence over Asian affairs. Contrary to some expectations, the end of colonial rule did not lead to a sudden cessation of Western business activity in Asia. As of 1970, 123 Indian companies were still owned by Britons, while the subcontinent remained an important destination for British exports even until the 1960s. Similarly, the Philippines' economy continued to rely heavily on American investment, which grew from US$149 million in 1950 to more than US$1 billion in 1973. Indeed, the world-wide Western economic stake actually increased immediately after the colonial period, reaching its peak, relatively speaking, around 1975. For instance, companies owned by Europeans and Americans accounted during the 1970s for 70 per cent of the world's copper output, 75 per cent of its nickle output, and 80 per cent of its aluminium output. Thus key economic decisions in Asia were—and are—being taken by companies whose policies are framed in foreign boardrooms. The interests of these foreign companies may not necessarily coincide with those of the host country, as demonstrated by the Lockheed scandal in Japan and the 1985 Bhopal disaster in India.[3] In addition, the West's dominance of lending bodies such as the World Bank and the International Monetary Fund gives it enormous bargaining power in regard to the Third World, which relies on such bodies to provide vital capital for development.

insidiously
treacherously

More **insidiously**, perhaps, the West also maintains a military presence in Asia. Due to the outbreak in 1948 of the Cold War, the end of colonialism *per se* saw a flurry of Western (and especially American) diplomatic and military activity whose rationale was neatly summed up by the 1954 Judd Mission to Indo-China:

> Its [Vietnam's] position makes it a strategic key to the rest of South East Asia. If Indo-China should fall, Thailand and Burma would be in extreme danger, Malaya, Singapore and even Indonesia would become vulnerable to the Communist power-drive.[3]

Obsessed by this 'domino theory', the United States and its allies sent troops to the aid of South Korea in 1950, Vietnam in 1961, and Cambodia in 1970. Ten years later, apparently learning nothing from the American humiliation in Vietnam, the Soviet Union went to the rescue of the Marxist government of Afghanistan, which was under siege from Islamic fundamentalists. Subsequently, the Western media self-righteously described the Soviets' intervention as a classic, old-fashioned exercise in Great Power imperialism.

In 1972 the Americans withdrew their forces from Vietnam and in 1989 the Russians pulled their troops out of Afghanistan. But others remain. Although the Cold War has ended and the two superpowers appear bent on disarmament, they continue to operate a worldwide network of bases backed up by orbiting spy satellites. Asia figures largely in this military chess-game. For its part the Soviet Union has warships stationed in South Yemen and at Cam Ranh Bay in Vietnam, while America has more than 147 000 servicemen based permanently in Asia and the north-western Pacific, 37 000 in South Korea alone. Especially in Japan the continuing American presence causes much resentment. In 1970 the fiercely patriotic author Yukio Mishima struck a sympathetic chord when he ritually disembowelled himself at the Tokyo headquarters of the Japan Self Defence Force in protest at the Army's loss of its once revered martial spirit.[4]

Whether it is termed 'peripheral dependency', 'neocolonialism' or 'gunship diplomacy', the phenomenon of continuing Western hegemony is an undeniable, if unpleasant, feature of the modern world. Despite outward appearances, the Westerners in Asia did not meekly pack their bags and go home. Many stayed on, ready and eager to compete commercially in the post-colonial market place. Likewise, most Western governments succeeded in maintaining close ties with their former colonies. The transfer of power was often accompanied by the signature of a mutual defence treaty guaranteeing the metropolitan power access to local bases. This happened in the cases of the Philippines, Ceylon and Malaya. It seems that, in the words of the Dutch historian H L Wesseling, 'we have to accept that [territorial] empire is one, but not the only form of imperialism, that there was imperialism before and after and even instead of empire'.[5]

An Era Of Asian Dominance?

In recent years, the pendulum of world power has been slowly swinging back Asia's way. The Soviet Union is in deep trouble economically, and the United States is less powerful than it once was, its industries contributing only 20 per cent of the world's gross national product in 1988 compared to 40 per cent immediately after World War II. At the same time, Western Europe has become more **insular** as a result of the formation of the European Economic Community (EEC) which greatly enlarged the 'domestic' market of the participating countries. As late as 1960, the bulk of the United Kingdom's trade was with its former colonies, but since its entry into the EEC, the pattern has changed dramatically. India used to absorb around 25 per cent of Britain's exports, but now accounts for only about 2 per cent. Similarly the great majority of British foreign investments (over 20 per cent) now ends up on the Continent. This, in turn, prompted the former colonies to look for new trading outlets in their own region: in 1967 Japan eclipsed the United Kingdom as Australia's major trading partner.

insular ignorant or indifferent to other countries and their cultures

It was a symbolic realignment. As the West has slipped, Japan has increasingly moved into the breach to the point where it now looms as a real threat to the economic hegemony of America. As early as 1970 Japan led the world in shipbuilding. Today it has the second biggest economy, eight of the top ten richest banks, the second and seventh biggest cities, is the world's major exporter of capital, and easily has the largest trade surplus of any developed country—US$86 billion in 1989.

As a consequence of prosperity, there came power and a new sense of national esteem. In 1953 opinion pollsters asked Japanese whether, as 'compared to Westerners', they saw themselves as superior or inferior. Only 20 per cent said the former. In 1983, the survey was repeated. This time, 53 per cent of those polled replied that they felt superior. Five years later, the racist implications of this poll were made explicit when Prime Minister Takashita offered mock sympathy to American GI Servicemen on Okinawa who, as a result of the appreciation of the yen against the dollar, could 'no longer afford the local bar girls'.[6] Similarly, Japanese executives these days generally believe that their own technology is at least as good, if not better, than that of the West. Even Japanese tourists, to judge from a recent Australian survey, are becoming more and more critical of the amenities and standards of service available in Western countries.[7]

Japan, of course, has been the leader in Asia's revival, but recently other Asian countries have also started to make their mark on the world stage. With its vast population and land area, China has always been, in some senses, a great power. It already boasts, after the Soviet Union, the world's largest standing army, with more than 3 million men and women under arms. Now, the modernisation policy initiated by Deng Xiaoping promises to raise China's economy to a level proportionate with its military capacity. Nevertheless, to date, China's economic achievements have been eclipsed by its smaller **Sinicised** neighbours: South Korea, Taiwan, Hong Kong and Singapore—the so-called 'Four Dragons'. During the 1970s and 1980s these countries grew at a rate not witnessed since the heyday of the Industrial Revolution in England: South Korea's gross national product expanded by an average of 8.3 per cent per year between 1962 and 1982. Significantly, the South Korean capital of Seoul has recently moved up from eighth to fifth in the 'top 10' of the world's cities. Last but not least there is India. Along with China, it has been a member of the exclusive nuclear club since 1974. India has the world's fourth largest army, a growing blue-water navy, 740 combat aircraft, and is in the process of producing its own intermediate-range ballistic missiles. Currently, its economy is the world's twelfth largest, but by the year 2000 its gross national product is expected to exceed that of France. Moreover, it is beginning to act like a Great Power, sending troops to Sri Lanka in 1987 and to the Maldives in 1988 and launching what amounted to a blockade of Nepal in 1989. Hypocritically, Western politicians and journalists have attacked this new posture. But the Indian strategist Krishnaswamy Subrahmanyam pointed out in an interview with *Time* magazine that it was both logical and inevitable:

Sinicised
controlled by China

> One out of every six people in the world is an Indian. In any democratic structure, India would have an effective say. But you in the West devised a world order in which the second largest country isn't even a permanent member of the U.N. Security Council. That's a big omission![8]

Having been subordinate for a long time, Asia is determined to recover its rightful place in world affairs.

An Asian resurgence has long been expected. As early as 1894, during the Sino-Japanese War, an English journalist wrote:

> The Oriental, with his power of retaining health under conditions under which no European could live, with his savage daring when

> roused, with his inborn cunning, lacks only the superior knowledge of civilisation to be the equal of the European in warfare as well as industry. Under the Japanese emperor the dream of the supremacy of the Yellow Race in Europe, Asia and even Africa . . . would be no longer mere nightmares.[9]

Today, the West's apprehensions are expressed rather differently, but their meaning is much the same. For example, the Pulitzer Prize-winning American author, Gore Vidal, wrote in the aftermath of the Wall Street crash of 1985:

> in the Autumn, the money power shifted from New York to Tokyo, and that was the end of our empire. Now the long-feared Asiatic colossus takes its turn as world leader, and we—the white race—have become the yellow man's burden. Let us hope he will treat us more kindly than we treated him.[10]

Particularly since the price of oil was increased dramatically in 1973 by the Organisation of Petroleum Exporting Countries, sentiments of this sort have proliferated. More and more, Westerners are starting to think, like Vidal, that their era of leadership might be over. Are they right? Will the historians of the 21st century look back on the 1980s as the beginning of an age of Asian dominance?

So far the statistics do not support this hypothesis. Despite the meteoric rise of Japan and the Four Dragons, the 'money power', to use Vidal's phrase, remains predominantly centred in the West. In 1982, Asia's total gross national product was US$1.68 million million, a little less than Western Europe's (US$1.70 million million) and substantially less than America's (US$2.14 million million). The gap is closing, but it will be well into the next century before the West, collectively, is overtaken. Moreover, the economic renaissance of Asia remains patchy and unbalanced. The gross national product per capita is the best simple measure we have of a country's standard of living. At US$12 850, Japan compares very favourably with the developed nations of Europe and North America, but the rest of Asia still lags far behind. For all its military might, the Peoples' Republic of China remains a poor country, yet its income per head of just US$310 is almost double that of disaster-prone Bangladesh, where the average daily food intake is a mere 8.6 kilojoules—far less than what the United Nations regards as necessary for survival. Japan, certainly, has joined the ranks of the superpowers. However, Asia as a whole still has much work to do before it can lay claim to world leadership. The East's turn will come—but currently European man and Western culture still reign supreme.

Additional Reading, Viewing and Notes

Chapter 1

Books

C R Boxer, *The Dutch Seaborne Empire 1600–1800*, Penguin Books, 1973.

J Clavell, *Shogun*, Dell Books, 1986.

H Furber, *Rival Empires of Trade in the Orient, 1600–1800*, University of Minnesota Press, 1975.

G V Scammell, *The First Imperial Age: European Overseas Expansion* c. *1400–1715*, Unwin Hyman, 1989.

Videos

Roads to Xanadu, ABC TV, Australia, 1989, episode 1.

The Flying Dutchmen, Veronica TV and Fuga Films, The Netherlands, 1986, episode 5.

Notes

1 Gomes Eanes de Zurara, quoted in D J Boorstin, *The Discoverers*, p. 166.
2 D J Boorstin, *The Discoverers*, p. 157.
3 A J Toynbee, *A Study of History, Vol. II*, Oxford, 1935, pp. 203–4.
4 H Delpar, *The Discoverers: an Encyclopedia of Explorers and Exploration*, p. 11.
5 C M Cipolla, *European Culture and Overseas Expansion*, p. 101.
6 J Morris, *Pax Britannica: The Climax of an Empire*, p. 119.
7 G M Young, *Early Victorian England*, c. *1830–1865*, pp. 104, 137.
8 Viscount Hardinge of Penshurst, *My Indian Years 1910–1916*, p. 1.

Chapter 2

Books

R F Betts, *The False Dawn: European Imperialism in the Nineteenth Century*, University of Minnesota Press, 1975.

J Clavell, *Taipan*, Dell Books, 1986.

C Hibbert, *The Dragon Wakes: China and the West, 1973–1911*, Penguin Books, 1984.

J Morris, *Heaven's Command: An Imperial Progress*, Penguin Books, 1979.

Notes

1 Quoted in *The British Empire, Vol. I*, p. 147.
2 Quoted in W Baumgart, *Imperialism: The Idea and Reality of British and French Colonial Expansion 1880–1914*, p. 39.
3 Sir John Seeley, *The Expansion of England*, p. 179.
4 J F Cady, *The Roots of French Imperialism in Eastern Asia*, p. 294.
5 K K Aziz, *The British in India: a Study in Imperialism*, p. 47.
6 Disraeli to Lord Malmesbury, 13 August 1852, quoted in C C Eldridge, *England's Mission: The Imperial Idea in the Age of Gladstone and Disraeli, 1868–1880*, p. 178.
7 It was J A Hobson who first cited 'the lust of the spectator' as a factor in European expansion. See J A Hobson, *Imperialism: A Study*, p. 215.

Chapter 3

Books

D R Headrick, 'The Tools of Imperialism: Technology and the Expansion of European Colonial Empires in the Nineteenth Century', in *Journal of Modern History*, Vol. 51, 1979, pp. 231–63.

R Kipling, *Life's Handicap*, Macmillan, 1982.

P J Marshall, 'Western Arms in Maritime Asia in the Early Phases of Expansion', in *Modern Asian Studies*, Vol. 14, 1980, pp. 13–28.

P Mason, *A Matter of Honour: An Account of the Indian Army, its Officers and Men*, Jonathan Cape, 1975.

Film

Fifty-five Days at Peking, USA, 1962, directed by Nicholas Ray.

Notes

1 Quoted in *The British Empire, Vol. V*, p. 608.
2 Quoted in *The British Empire, Vol. 2*, p. 532.
3 T Smith, *The Pattern of Imperialism: The United States, Great Britain and the Late-Industrialising World Since 1815*, C.U.P., p. 22.
4 V G Scammel, *The World Encompassed: the First European Maritime Empires, c. 800–1650*, p. 265.
5 P Woodruff (pseud.), *The Men Who Ruled India, Vol. II, The Guardians.*
6 Sir John Seeley, *The Expansion of England*, p. 203.
7 L Woolf, *An Autobiography*, O.U.P., p. 158.

Chapter 4

Books

V G Kiernan, *Lords of Human Kind: European Attitudes to the Outside World in the Imperial Age*, Penguin Books, 1972.
D A Low, *Lion Rampant: Essays in the Study of British Imperialism*, Frank Cass and Co., 1973.
R von Albertini, *European Colonial Rule 1880–1940*, Clio Press, 1982.

Film

Films From the Raj, Cambridge South Asian Studies Centre, 1974 (home movies taken by British officers and their families in the early 20th century).

Videos

Journey Into India, ABC, 1978, episode 3: 'Kipling's India'.
The Triumph of the West, BBC, 1984, episode 10: 'The Confident Aggressors'.

Notes

1 Minute by Lord William Bentinck, Governor-General of India, dated 14 October 1833, quoted in G D Bearce, *British Attitudes towards India*, p. 162.
2 M Osborne, *The French Presence in Cochinchina and Cambodia: Rule and Response (1859–1905)*, p. 38.
3 Mayo to Lord Argyll, 1 September 1871, Argyll Papers, India Office Library, Vol. 4.
4 S Sarkar, *Modern India, 1885–1947*, p. 1.
5 D K Fieldhouse, *The Colonial Empires: a Comparative Survey From the Eighteenth Century*, p. 381.

Chapter 5

Books

J Masters, *Bhowani Junction*, Sphere Books, 1985.
R Robinson, 'Non-European Foundations of Imperialism: Sketch For a Theory of Collaboration' in W R Louis (Ed), *Imperialism: The Robinson and Gallagher Controversy*, New Viewpoints, 1976.
P Tandon, *Punjabi Century, 1857–1947*, California University Press, 1968.
P Toer, *This Earth of Mankind*, Penguin Books, 1975.

Video

A Faded Glory, ABC, 1974 (on the ruler and the state of Bikaner in Rajasthan).

Notes

1 D A Low, 'Lion Rampant', in *Journal of Commonwealth Political Studies*, Vol. II, 1963, p. 235.
2 D A Low 'Lion Rampant', in *Journal of Commonwealth Political Studies*, Vol. II, 1963, p. 246.
3 *Thug* is an Indian word, and derives from the name of a fearsome cult, based on the worship of the goddess Kali, which operated in north India. The *thugs* preyed upon the highways, robbing stray travellers and strangling them with a yellow scarf. It is estimated that as many as 2 million people may have been murdered in this way between 1600 and 1850, by which time the British had succeeded in stamping out the practice.
4 Abdullah thought he was on a good thing but found, with experience, that British protection left him little control over his State. Accordingly he arranged for the murder of the British Resident, J W W Birch, which led to his deposition and exile in 1875.

Chapter 6

Books

J Ch'en, *China and the West*, Hutchinson, 1979.
J B Cowley (Ed), *Modern East Asia: Essays in Interpretation*, Harcourt Brace, 1970.
M N Srinivas, *Social Change in Modern India*, University of California Press, 1973.

Videos

Roads to Xanadu, ABC, 1989, episodes 2–4.
The Triumph of the West, BBC, 1984, episode 11.

Notes

1 Quoted in P G Kontos *et al*, *Patterns of Civilization: Asia*, p. 239.
2 Quoted in J M Roberts, *The Triumph of the West*, p. 392.
3 R Murphey, *The Outsiders: the Western Experience in India and China*, p. 226.
4 China, for example, was required, under the Convention of 1860 (concluded after the Second 'Opium War') to allow foreign diplomats to reside permanently at the capital.

Chapter 7

Books

R Emerson, *From Empire to Nation: The Rise to Self-Assertion of Asian and African Peoples*, Beacon Press, 1960.
R K Narayan, *Waiting for the Mahatma*, Indian Thought Publications, 1978.
A Seal, *The Emergence of Indian Nationalism: Competition and Collaboration in the Later Nineteenth Century*, Cambridge University Press, 1968.

Film

Gandhi's India (BBC, 1969).

Video

Struggle For China: The Story of the Chinese Revolution 1900–1949 (Yorkshire TV, 196–).

Notes

1 H B Lamb, *Vietnam's Will to Live: Resistance to Foreign Aggression From Early Times Through the Nineteenth Century*, p. 296.
2 The Arya Samaj was, in some respects, strongly influenced by the Protestant missionary movement; but its basic thrust was revivalist in that it took its inspiration from the four oldest Hindu texts known as the *Vedas* and regarded all later writings as corrupt. The Ganapati Festival honoured the Hindu god of good fortune—the elephant-headed Ganesh. An old festival, it had almost died out when it was revived by Tilak and others in the 1890s. The Shivaji Festival, on the other hand, was an innovation of Tilak.
3 Lord Curzon infuriated Bengali Hindus by deciding in 1905 to partition their province in order to give the Muslim-majority in the east a measure of local autonomy. From 1905 to 1908, Bengal was wrecked by riots and boycotts on a scale never seen before in India.

Chapter 8

Books

L Collins & D Lapierre, *Freedom at Midnight*, Pan Books, 1977.
R Jeffrey (Ed), *Asia: The Winning of Independence*, Macmillan Ltd, 1981.
M Malgaonkar, *A Bend in the Ganges*, Orient Paperbacks, 1964.
T Smith (Ed), *The End of European Empire: Decolonization After World War II*, D C Heath & Co., 1975.

Videos

The End of Empire, Channel 4 TV/Granada TV, 1984, episodes 1–4.
Vietnam: A Television History, PBS Boston, 1982, episode 1.

Notes

1 Sir John Seeley, *The Expansion of England*, p. 196.
2 Quoted in J Bastin (Ed), *The Emergence of Modern Southeast Asia: 1511–1957*, p. 111.
3 Statement of 20 August 1917. Quoted in C H Philips (Ed), *The Evolution of India and Pakistan 1858 to 1947: Select Documents*, p. 264.
4 Quoted in L H Palmier, *Indonesia and the Dutch*, p. 35.
5 Sir Percival Griffiths, *The British Impact on India*, p. 356.
6 R Robinson, 'The Excentric Idea of Imperialism, With or Without Empire', in W J Mommsen & J Osterhamel (Eds), *Imperialism and After: Continuities and Discontinuities*, p. 272.
7 E P Hoyt, *Japan's War: the Great Pacific Conspiracy*, p. 419.
8 N H Webster, *The Surrender of an Empire*, 1931, p. 314.
9 Quoted in L Collins and D Lapierre, *Freedom at Midnight*, p. 71.

Chapter 9

Books

F Darling, *The Westernization of Asia: A Comparative Political Analysis*, Schenkman Publishing Company, 1980.
D K Fieldhouse, *Colonialism 1870–1945: An Introduction*, Weidenfeld and Nicolson, 1981.
D K Fieldhouse, *The Colonial Empires: A Comparative Survey From the Eighteenth Century*, Weidenfeld and Nicolson, 1966.
I R Sinai, *The Challenge of Modernization: The West's Impact on the Non-Western World*, W W Norton & Co., 1964.

Video

The Flying Dutchmen, Episode 7.

Notes

1 Speech to the Royal Colonial Institute, 31 March 1897. In W D Handcock (Ed), *English Historical Documents 1874–1914, Vol. XII*, p. 389.
2 A Sen, *The State, Industrialization and Class Formations in India*, p. 69.
3 G C Allen & G Donnithorne, *Western Enterprise in Indonesia and Malaya: A Study in Economic Development*, p. 265.
4 I R Sinai, *The Challenge of Modernisation: The West's Impact on the Non-Western World*, p. 54.
5 P Marshall, *East India Fortunes: the British in Bengal*, p. 255.
See also H Furber, *Rival Empires of Trade in the Orient 1600–1800*, p. 334.

Epilogue

Notes

1 H Luthy, 'Colonization and the Making of Mankind' in G H Nadel & P Curtis (Eds), *Imperialism and Colonialism*, p. 36.
2 In 1979 Japan was rocked by the revelation that dozens of senior businessmen and politicians had been bribed by the American Lockheed Corporation in a bid to promote the sale of its Tristar aircraft. The list of accused was headed by the then Prime Minister, Tanaka, who was charged with pocketing 500 million yen (US$2 million).
In December 1984, 3150 people died and some 30 000 were permanently disabled when a Union Carbide chemical plant in the central Indian city of Bhopal exploded. An enquiry uncovered evidence of wilful negligence. In 1989 the Government of India accepted Union Carbide's offer of A$601.6 million compensation. However, the defeat of the Congress in the elections of 1989 has thrown the issue open again, and the new government of V P Singh is vowing to pursue criminal charges and fight for the A$3.5 billion originally sought from the American company.
3 Quoted in M Gurtov, *The First Vietnam Crisis: Chinese Communist Strategy and United States Involvement 1953–4*, p. 26.
4 Recently, the Soviet Government has started withdrawing ships and aircraft from Cam Ranh Bay, and Foreign Minister Shevardnadze has been quoted as saying that 'The day is near when there will be no Soviet military presence in Asia beyond Soviet borders' (The *Age*, 20 January 1990). So far there has been no sign of the Americans reciprocating.
5 H L Wesseling, 'Imperialism and Empire: an Introduction', in W J Mommsen and J Osterhammel (Eds), *Imperialism and After: Continuities and Discontinuities*, 1986, p. 8.
6 Quoted in The *Age*, 27 July 1988.
7 According to The *Age*, 10 January 1990, more than half of the Japanese interviewed said they were unhappy with shop trading hours, 21 per cent (compared to 11 per cent in 1986) said they were dissatisfied with airport facilities, and 17 per cent (compared to 12 per cent in 1986) complained about service in hotels and restaurants.
8 H Munro, 'Superpower Rising', *Time*, 3 April 1989, pp. 17–18.
9 Quoted in A Iriye, 'Imperialism in East Asia', in J B Crowley (Ed), *Modern East Asia: Essays in Interpretation*, p. 145.
10 G Vidal 'The Day the U.S. Empire Died', The *Age Extra*, 2 March 1986.

Bibliography

Adas, Michael. *The Burma Delta: Economic Development and Social Change on an Asian Rice Frontier, 1852–1941.* Wisconsin: University of Wisconsin Press, 1974.

Ahmad Khan, Sir Sayeed. *History of the Bijnor Rebellion.* East Lansing (MI): Asian Studies Center, Michigan State University, n.d.

Ahmad Khan, Sir Sayeed. *The Cause of the Indian Revolt.* East Lansing (MI): Asian Studies Center, Michigan State University, 1873.

Alavi, Hamza & Harriss, John (Eds). *Sociology of 'Developing Societies': South Asia.* Houndmills (UK): Macmillan Co, 1989.

Allen, G C & Donnithorne, Audrey C. *Western Enterprise in Indonesia and Malaya.* London: George Allen & Unwin Ltd, 1954 (rpt 1962).

Avineri, Shlomo (Ed). *Karl Marx on Colonialism and Modernization.* Garden City (NY): Anchor Books Doubleday & Co Inc, 1969.

Aziz, K K. *The British in India: A Study in Imperialism.* Islamabad: National Commission on Historical and Cultural Research, 1976.

Bairoch, Paul. 'International Industrial Levels from 1750–1850' in *The Journal of European Economic History,* Vol 11, 1982.

Bairoch, Paul. 'The Main Trends in National Economic Disparities since the Industrial Revolution' in Paul Bairoch & Maurice Lévy-Leboyer (Eds), *Disparities in Economic Development since the Industrial Revolution.* London: The Macmillan Press Ltd, 1981.

Banerjea, Sir Surendranath. *A Nation in Making: Being the Reminiscences of Fifty Years of Public Life.* Bombay: Oxford University Press, 1925 (rpt 1963).

Barrett, David B (Ed). *World Christian Encyclopedia.* Nairobi: Oxford University Press, 1982.

Bastin, John (Ed). *The Emergence of Modern Southeast Asia: 1511–1957.* Englewood Cliffs (NJ): Prentice-Hall Inc, 1967.

Baum, Vicki. *A Tale from Bali.* London: Geoffrey Bles, 1937.

Baumgart, Winfried. *Imperialism: The Idea and Reality of British and French Colonial Expansion, 1880–1914.* Oxford: Oxford University Press, 1982 (rpt 1986).

Beames, John. *Memoirs of a Bengal Civilian.* Columbia: Chatto & Windus South Asia Books, 1961 (rpt 1984).

Bearce, George D. *British Attitudes Towards India 1784–1858.* Oxford: Oxford University Press, 1961.

Benda, Harry J & McVey, Ruth T (Eds). *The Communist Uprisings of 1926–1927 in Indonesia: Key Documents.* Ithaca (NY): Modern Indonesia Project, Southeast Asia Program, Department of Far Eastern Studies, Cornell University, 1960.

Bennett, George (Ed). *The Concept of Empire, Burke to Attlee, 1774–1947.* London: Adam & Charles Black, 1953 (rev 1967).

Blair, Emma Helen & Robertson, James Alexander (Eds). *The Philippine Islands 1493–1898, Vol. XLVI, 1727–1739.* Mandaluyong, Rizal: Cachos Hermanos Inc, 1973.

Blunt, Wilfrid Scawen. *My Diaries, Vol 1.* London: Martin Secker, 1919–1920.

Blyn George. *Agricultural Trends in India, 1891–1947: Output, Availability, and Productivity.* Philadelphia: University of Philadelphia Press, 1966.

Boorstin, Daniel J. *The Discoverers* (1983). Harmondsworth: Penguin Books, 1986.

Bosworth Smith, R. *Life of Lord Lawrence,* 2 vols. London: Smith, Elder & Co, 1883.

Bowring, Sir John. *The Kingdom and People of Siam, Vol. I.* London: John W Parker and Son, 1857.

Boxer, C R. *The Christian Century in Japan, 1549–1650.* Berkeley: University of California Press, 1951.

Boxer, C R. *The Dutch Seaborne Expire: 1600–1800* (1965). London: Penguin Books, by arrangement with Hutchinson of London, 1988.

Bright, John & Rogers, James E Thorold (Eds). *Speeches on Questions of Public Policy by Richard Cobden M. P.* London: Macmillan & Co, 1878.

Brown, Michael Barratt. *The Economics of Imperialism*. Harmondsworth: Penguin Books, 1974 (rpt 1976).
Buttinger, Joseph. *Vietnam: A Dragon Embattled. Vol. 1, From Colonialism to the Vietminh*. New York: Frederick A Praeger, 1967.
Cady, John F. *The Roots of French Imperialism in Eastern Asia*. Ithaca (NY): Cornell University Press, 1954 (rev 1967).
Caldwell, Malcolm & Utrecht, Ernst. *Indonesia: An Alternative History*. Sydney: Alternative Publishing Co-operative Ltd, 1979.
Cambridge Economic History of India, Vol 2. Cambridge University Press.
Carroll, Lucy. 'The Seavoyage Controversy and the Kayasthas of North India 1901–1909' in *Modern Asian Studies*, Vol 13, 1979.
Chang, Queeny. *Memories of a Nonya*. Singapore: Eastern Universities Press, 1981.
Chaudhuri, Nirad C. *The Autobiography of an Unknown Indian* (1951). Bombay: Jaico Publishing House, 1966.
Chow, Tse-tsung. *The May Fourth Movement: Intellectual Revolution in Modern China* (1960). Stanford (CA): Stanford University Press, 1967.
Cipolla, Carlo M. *European Culture and Overseas Expansion*. Harmondsworth: Penguin Books, 1965 (rpt 1970).
Collins, Larry & Lapierre, Dominique. *Freedom at Midnight*. London: William Collins Sons & Co Ltd, 1975.
Collis, Maurice. *Raffles* (1966). Singapore: Graham Brash Pte Ltd, 1986.
Colomb, Brig-Gen G H. 'India Without Britain' in *The National Review*, Vol LXXIV,1924–25.
Conrad, Joseph. *Lord Jim: A Tale*. London: J M Dent & Sons Ltd, 1968.
Crowley, James B (Ed). *Modern East Asia: Essays in Interpretation*. New York: Harcourt, Brace & World Inc, 1970.
Curtin, Philip D (Ed). *Imperialism*. New York: Walker & Co, 1972.
Darling, Frank. *The Westernization of Asia*. Schenkman Publishing Co, 1980.
Darling, Sir Malcolm. *Apprentice to Power: India, 1904–1908*. London: The Hogarth Press, 1966.
de Bary, William T (Ed). *Sources of Chinese Tradition*. New York: Columbia University Press, 1960.
de Bary, William T (Ed). *Sources of Indian Tradition, Vol. 2*. New York: Columbia University Press, 1958.
de Bary, William T (Ed). *Sources of Japanese Tradition*. New York: Columbia University Press, 1958.
Delpar, Helen (Ed). *The Discoverers: An Encyclopedia of Explorers and Exploration*. New York: McGraw Hill Book Co, 1980.
Dilke, Sir Charles Wentworth. *Greater Britain: A Record of Travel in English-Speaking Countries During 1866 and 1867*. London: Macmillan & Co, 1872.
Eldridge, C C. *England's Mission: The Imperial Idea in the Age of Gladstone and Disraeli 1868–1880*. London: Macmillan & Co, 1973.
Fanon, Franz. *The Wretched of the Earth*. New York: Grove Press, 1963.
Feith, Herbert (Ed). *Indonesian Nationalism and Revolution: Six First-hand Accounts*. Clayton: Monash University, 1969 (rpt 1971).
Fieldhouse, D K. *The Colonial Empires: A Comparative Survey from the Eighteenth Century*. London: Weidenfeld & Nicolson, 1966.
Forbes-Mitchell, William. *Reminiscences of the Great Mutiny 1857–59*. London: Macmillan & Co Ltd, 1893 (rpt 1910).
Francisco, Mariel N & Arriola, Fe Maria C. *The History of the Burgis*. Quezon City: GCF Books, 1987.
Franke, Wolfgang. *China and the West* (transl R A Wilson; 1967). Columbia: University of South Carolina Press, 1968.
Fraser, Sir Andrew H L. *Among Indian Rajahs and Ryots*. Allahabad: Chugh Publications, 1975.
Fukuzawa, Yukichi. *The Autobiography of Fukuzawa Yukichi*. Tokyo: The Hokuseido Press, 1960.
Furber, Holden. *Rival Empires of Trade in the Orient 1600–1800*. Minneapolis: University of Minnesota Press, 1976.
Gandhi, M K. *Collected Works, Vol. XVIII*. Ahmedabad: The Publications Division, Ministry of Information and Broadcasting, Government of India, 1965.
Gettleman, Marvin E (Ed). *Vietnam: History, Documents, and Opinions on a Major World Crisis* (1965). Harmondsworth: Penguin Books, 1967.
Ghosh, Suresh Chandra. *The Social Condition of the British Community in Bengal 1757–1800*. Leiden: E J Brill, 1970.
Gibson, Walter M. *The Prison of Weltevreden; and a Glance at the East Indian Archipelago*. New York: J C Riker, 1855.
Goonatilake, Susanthra. *Crippled Minds: An Exploration into Colonial Culture*. Delhi: Vikas Publishing House Pvt Ltd, 1982.
Griffiths, Sir Percival. *The British Impact on India*. London: Frank Cass & Co Ltd, 1952 (rpt 1965).
Gunther, John. *Inside Asia*. London: Hamish Hamilton, 1939.
Gurtov, Melvin. *The First Vietnam Crisis: Chinese Communist Strategy and United States Involvement, 1953–1954*. New York: Columbia University Press, 1967.
Haga, B J. 'Influence of the Western Administration on the Native Community in the Outer Provinces'

in B Schreike (Ed), *The Effect of Western Influence of Native Civilization in the Malay Archipelago*. Batavia: G Kolff & Co, 1929.

Hakluyt, Richard. *Voyages and Discoveries: The Principal Navigations Voyages, Traffiques and Discoveries of the English Nation* (Ed Jack Beeching). Harmondsworth: Penguin Books, 1972 (rpt 1985).

Hale, H W. *Political Trouble in India 1917–1937*. Allahabad: Chugh Publications, 1974.

Handcock, W D (Ed). *English Historical Documents 1874–1914*. London: Eyre & Spottiswoode, 1977.

Hardinge, Lord, of Penshurst. *My Indian Years 1910–1916*. London: John Murray, 1948.

Harrison, John A. *China since 1800*. New York: Harcourt, Brace & World Inc, 1967.

Headrick, D R. 'The Tools of Imperialism: Technology and the Expansion of European Colonial Empires in the Nineteenth Century' in *Journal of Modern History*, Vol 51, 1979.

Hibbert, Christopher. *The Dragon Wakes: China and the West, 1793–1911* (1970). Harmondsworth: Penguin Books, 1984.

Hill, S C (Ed). *Indian Records Series, Bengal in 1756–1757*, Vol III. New York: AMS Press, 1968.

Hobsbawm, E J. *The Pelican Economic History of Britain. Vol. 3: From 1750 to the Present Day: Industry and Empire* (1968). Harmondsworth: Penguin Books, 1972.

Hobson, J A. *Imperialism: A Study*. London: Unwin Hyman, 1902 (3rd edn 1938; rpt 1988).

House Documents, 55th Congress, No 3, 1898. United States Congress.

Hoyt, Edwin P. *Japan's War: The Great Pacific Conflict 1853–1952*. London: Century Hutchinson, 1986 (rev 1987).

Hull, Cordell. *The Memoirs of Cordell Hull, Vol. I*. London: Hodder & Stoughton, 1948.

Hunter, Lord. *Disorders Inquiry Committee 1919–1920: Report*. Calcutta: Superintendent Government Printing, 1920.

Indian Mirror, 20 May 1883.

Iyer, Raghavan (Ed). *The Moral and Political Writings of Mahatma Gandhi Vol I: Civilization, Politics and Religion*. Oxford: Oxford University Press, 1986.

Jeffrey, Robin (Ed). *Asia—The Winning of Independence*. London: The Macmillan Press Ltd, 1981.

Keene, Donald. *The Japanese Discovery of Europe, 1720–1830*. Stanford (CA): Stanford University Press, 1952 (rev 1969).

Ker, James Campbell. *Political Trouble in India 1907–1917*. Daryaganj, Delhi: Oriental Publishers, 1917 (rpt 1973).

Kipling, Rudyard. *Rudyard Kipling's Verse*. London: Hodder & Stoughton, 1940.

Kish, George (Ed). *A Source Book in Geography*. Cambridge (MA): Harvard University Press, 1978.

Klein, Ira. 'Malaria and Mortality in Bengal, 1840–1921' in *Indian Economic and Social History Review*, Vol IX, 1972.

Kontos, Peter G *et al.* (Eds). *Patterns of Civilization: Asia*. New York: Cambridge Book Co, 1974.

Kratoska, Paul H (Ed). *Honourable Intentions*. Singapore: Oxford University Press, 1983.

Lamb, Helen B. *Vietnam's Will to Live: Resistance to Foreign Aggression from Early Times Through the Nineteenth Century*. New York: Monthly Review Press, 1972.

Langer, William L. *The Diplomacy of Imperialism 1890–1902*. New York: Alfred P Knopf, 1935 (rev 1960).

Lawrence, Sir Walter Roper. *The India We Served*. London: Cassell & Co Ltd, 1928.

Lenin, V I. *Imperialism: The Highest Stage of Capitalism*. Berlin, 1916.

Ley, Charles David (Ed). *Portuguese Voyages 1498–1663*. London: J M Dent & Sons Ltd, 1947 (rpt 1960).

Lo, Hui-Min (Ed). *The Correspondence of G. E. Morrison. Vol. I. 1895–1912*. Cambridge: Cambridge University Press, 1976.

Louis, W Roger (Ed). *The Robinson and Gallagher Controversy*. New York: New Viewpoints, 1976.

Low D A. *Lion Rampant: Essays in the Study of British Imperialism*. London: Frank Cass & Co Ltd, 1973 (rpt 1974).

Lunt, James (Ed). *From Sepoy to Subedar being the Life and Adventures of Subedar Sita Ram, a Native Officer of the Bengal Army written and related by himself* (1873). Delhi: Vikas Publications, 1970.

Mansergh, Nicholas (Ed). *The Transfer of Power, 1942–7, Vol 7*. London: HMSO, 1977.

Marshall, P J. *East Indian Fortunes: The British in Bengal in the Eighteenth Century*. Oxford: Oxford University Press 1976.

Marshall, P J. *Problems of Empire: Britain and India 1757–1813*. London: George Allen & Unwin Ltd, 1968.

Martin, W A P. *A Cycle of Cathay or China, South and North*. New York: Paragon Book Gallery Ltd, 1900.

Mason, Philip. *A Matter of Honour*. New York: Holt, Rinehart & Winston, 1974.

Masters, John. *Bhowani Junction* (1954). London: Sphere Books Ltd, 1983.

Maugham, W Somerset. *Collected Short Stories, Volume 4*. London: Pan Books Ltd/William Heinemann Ltd, 1951 (rpt 1981).

McIntyre, W David. *The Imperial Frontier in the Tropics, 1865–75*. London: Macmillan & Co Ltd, 1967.

McLane, John R (Ed). *The Political Awakening in*

India. Englewood Cliffs (NJ): Prentice-Hall Inc, 1970.

Mehta, Ved. *Daddyji*. London: Secker & Warburg, 1972.

Mills, Lennox A. *Ceylon Under British Rule 1795–1932: With an account of the East India Company's Embassies to Kandy 1762–1795* (1933). London: Frank Cass & Co Ltd, 1964.

Moffat, Abbot Low. *Mongkut, the King of Siam*. Ithaca (NY): Cornell University Press, 1961 (rpt 1968).

Mommsen, Wolfgang J & Osterhammel, Jurgen (Eds). *Imperialism and After: Continuities and Discontinuities*. London: Allen & Unwin, 1986.

Montagu, E S & Chelmsford, Lord. *Report on Indian Constitutional Reforms*. Calcutta: Superintendent Government Printing, 1918.

Moon, Penderel (Ed). *Wavell: The Viceroy's Journal*. London: Oxford University Press, 1973.

Morris, James. *Pax Britannica: Climax of an Empire*. London: Faber & Faber, 1968.

Morris, James. *Heaven's Command: An Imperial Progress* (1973). Harmondsworth: Penguin Books Ltd, 1980.

Morrison, George Ernest. *An Australian in China: Being the Narrative of a Quiet Journey Across China to Burma*. London: Horace Cox, 1895.

Muir, Ramsay (Ed). *The Making of British India 1756–1858*. Lahore: Oxford University Press, 1915 (rpt 1969).

Mundy, Captain Rodney. *Narrative of Events in Borneo and Celebes down to the Occupation of Labuan, 2 Volumes*. London: John Murray, 1848.

Murphey, Rhoads. *The Outsiders: The Western Experience in India and China*. Ann Arbor: The University of Michigan Press, 1977.

Murray Martin J. *The Development of Capitalism in Colonial Indochina (1870–1940)*. Berkeley (CA): University of California Press, 1980.

Nadel, George H & Curtis, Perry (Eds). *Imperialism and Colonialism*. New York: The Macmillan Co, 1964 (rpt 1966).

National Archives of India, Revenue Proceedings, A Series, June 1867, No 30.

Nehru, Jawaharlal. *An Autobiography, With Musings on Recent Events in India* (1936). Bombay: Allied Publishers Pte Ltd, 1962.

Newton, Arthur Percival (Ed). *Travel and Travellers of the Middle Ages*. London: Routledge and Kegan Paul, 1926 (rev 1968).

Nock, O S. *Railways of the World, Vol. 5, Railways of Asia and the Far East*. London: Adam and Charles Black, 1974.

Norman, Dorothy (Ed). *Nehru: The First Sixty Years, Vol 1*. London: Bodley Head, 1965.

O'Dwyer, Sir Michael. *India as I Knew It 1885–1925*. London: Constable & Co Ltd, 1925.

Orwell, George. *Burmese Days* (1934). Harmondsworth: Penguin Books Ltd, 1975.

Osborne, Milton E. *The French Presence in Cochinchina and Cambodia: Rule and Response (1859–1905)*. Ithaca (NY): Cornell University Press, 1969.

O'Toole, G J A. *The Spanish War: An American Epic—1898*. New York: W W Norton & Co, 1984.

Palmier, Leslie H. *Indonesia and the Dutch*. London: Oxford University Press, 1962.

Panikkar, K M. *Asia and Western Dominance: A Survey of the Vasco Da Gama Epoch of Asian History 1498–1945*. London: George Allen & Unwin Ltd, 1953 (rev 1974).

Parkinson, C Northcote. *British Intervention in Malaya 1867–1877*. Singapore: University of Malaya Press, 1960.

Parliamentary Debates, 3rd series, Vol XIX. Great Britain.

Parliamentary Debates, 4th series, Vol XXX. Great Britain.

Paton, William. *Alexander Duff: Pioneer of Missionary Education*. London: Student Christian Movement, 1923.

Penders, C L M (Ed). *Indonesia: Selected Documents on Colonialism and Nationalism, 1830–1942*. St Lucia (Qld): University of Queensland Press, 1977.

Philips, C H. *The Evolution of India and Pakistan 1858 to 1947: Select Documents*. London: Oxford University Press, 1962 (rpt 1964).

Postgate, R. 'Echoes of a Revolt' in *The New Republic*, 22 May 1935.

Public Record Office, CAB 66/26. Great Britain.

Rajadhon, Phya Anuman. *Life and Ritual in Old Siam: Three Case Studies of Thai Life and Customs*. New Haven (CT): Human Relations Area Files Press, 1961.

Rawlinson, H G. *The British Achievement in India: A Survey*. London: William Hodge & Co Ltd, 1948

Richards, Philip Ernest. *Indian Dust: Being Letters from the Punjab*. London: George Allen & Unwin Ltd, 1932.

Roberts, J M. *The Triumph of the West*. London: British Broadcasting Corporation, 1985.

Robertson, David. *The Penguin Dictionary of Politics*. London: Penguin Books Ltd, 1985 (rpt 1988).

Robinson, Ronald, Gallagher, John & Denny, Alice. *Africa and the Victorians: The Official Mind of Imperialism*. London: Macmillan & Co Ltd, 1961 (rpt 1963).

Rose, Kenneth. *Curzon: A Most Superior Person: A Biography* (1969). London: Macmillan Publishers Ltd, 1985.

Scammell, G V. *The World Encompassed: The First European Maritime Empires c. 800–1650*. Berkeley (CA): University of California Press, 1981.

Schumpeter, J A. 'Imperialism and Social Classes' in Harrison M Wright (Ed), *The 'New imperialism': Analysis of Late Nineteenth Century Expansion*. Boston, 1961.

eal, Anil. *The Emergence of Indian Nationalism: Competition and Collaboration in the Later Nineteenth Century.* Cambridge: Cambridge University Press, 1968.

eeley, J R. *The Expansion of England: Two Courses of Lectures.* London: Macmillan & Co Ltd, 1902.

en, Anupam. *The State, Industrialization and Class Formations in India: A neo-Marxist perspective on colonialism, underdevelopment and development.* London: Routledge & Kegan Paul, 1982.

ffin, William J. *The Thai Bureaucracy: Institutional Change and Development.* Honolulu: East-West Center Press, 1966.

mmons, Colin. '"De-industrialization", Industrialization and the Indian Economy c.1850–1947' in *Modern Asian Studies*, Vol 19, 1985.

mmons, Colin. 'Labour and Capital in the Gold Mining Industry of India'. Paper presented at the International Mining History Conference, Bochum, FRG, September 1989.

nai, I R. *The Challenge of Modernisation: The West's Impact on the Non-Western World.* New York: W W Norton & Co Inc, 1964.

ahrir, Soetan. *Out of Exile.* New York: The John Day Co, 1949.

nith, Tony (Ed). *The End of the European Empire Decolonization after World War II.* Lexington (MA): D C Heath & Co, 1975.

nith, Tony. *The Pattern of Imperialism: The United States, Great Britain, and the Late Industrializing World Since 1815.* Cambridge: Cambridge University Press, 1981.

ear, Percival. *The Nabobs A Study of the Social Life of the English in Eighteenth Century India.* London: Oxford University Press, 1963.

ephen, J F. 'Foundations of the Government of India' in *The Nineteenth Century*, Vol XIV, 1883.

ickney, Joseph L. *War in the Philippines and Life and Glorious Deeds of Admiral Dewey.* Chicago: Monarch, 1899.

ate, D J M. *The Making of Modern South-East Asia, Vol 2: The Western Impact: Economic and Social Change* (1954). Kuala Lumpur: Oxford University Press, 1979.

eng, Ssu-Yu & Fairbank, John K., & others (Eds). *China's Response to the West: A Documentary Survey 1839–1923.* New York: Atheneum, 1975.

he *Age*, 27 July 1988.

he *Age*, 20 January 1990.

he *Age Extra*, 2 March 1986.

he *Age Extra*, 28 November 1987.

he *Graphic*, 26 June 1897.

he *Illustrated London News*, 4 July 1857.

hompson, Virginia. *French Indochina.* New York: The Macmillan Co, 1937.

ime, 3 April 1989.

Toer, Pramoedya Ananta. *This Earth of Mankind: a novel* (1975). Ringwood (Vic): Penguin Books, 1982.

Tomlinson, B R. *The Political Economy of the Raj 1914–1947: The Economics of Decolonization in India.* London: The Macmillan Press Ltd, 1979.

Toynbee, Arnold J. *A Study of History, Vol II.* London: Oxford University Press, 1934 (rpt 1939).

Trotter, Captain Lionel J. *The Life of John Nicholson: Soldier and Administrator.* London: John Murray, 1897

Truong, Buu Lam. *Patterns of Vietnamese Response to Foreign Intervention: 1858–1900.* New Haven: Monograph Series No 11, Southeast Asia Studies, Yale University, 1967.

Tung, William L. *China and the Foreign Powers: The Impact of and Reaction to Unequal Treaties.* Dobbs Ferry (NY): Oceana Publications Inc, 1970.

von Albertini, Rudolf. *Decolonization: The Administration and Future of the Colonies 1919–1960.* Garden City (NY): Doubleday & Co Inc, 1971.

von Albertini, Rudolf. *European Colonial Rule, 1880–1940: The Impact of the West on India, Southeast Asia, and Africa.* Oxford: Clio Press, 1982.

Ward, A B. *Rajah's Servant.* Ithaca (NY): Southeast Asia Program, Department of Asian Studies, Cornell University, 1966.

Webster, Nesta H. *The Surrender of an Empire.* London: Boswell Printing & Publishing Co Ltd, 1931.

Whitcombe, Elizabeth. *Agrarian Conditions in Northern India, Vol I: The United Provinces Under British Rule, 1860–1900.* Berkeley (CA): University of California Press, 1972.

Wilkinson, Endymion. *Japan Versus Europe: A History of Misunderstanding.* Harmondsworth: Penguin Books, 1983.

Wilson, Philip Whitwell (Ed). *Greville Diary, Vol. I.* London: William Heinemann Ltd, 1927.

Wolpert, Stanley A. *Tilak and Gokhale: Revolution and Reform in the Making of Modern India.* Berkeley: University of California Press, 1962.

Wong, Lin Ken. 'Singapore: Its Growth as an Entrepot Port, 1819–1914' in *Journal of South East Asian Studies*, Vol 9, 1978.

Woodruff, Philip. *The Men Who Ruled India: The Guardians.* London: Jonathan Cape, 1954 (rpt 1963).

Woolf, Leonard. *An Autobiography, Vol 1: 1880–1911* (1960). Oxford: Oxford University Press, 1980.

Young, G M (Ed). *Early Victorian England 1830–1865, Vol I.* London: Oxford University Press, 1934 (rpt 1951).

Index

Acknowledgements

The author and publishers are grateful to copyright holders for permission to reproduce copyright material. Sources of all textual extracts are listed in detail in the Bibliography; the following list comprises current holders of copyright as indicated by imprint pages:

A & C Black, *Railways of Asia and the Far East* by O S Nock; *The Concept of Empire: Burke to Attlee 1774–1947* by George Bennett.
The *Age*, extract from 'Ooty: Living echo of a dead empire' by Geoffrey Kenihan.
Alternative Publishing Co-operative Ltd, *Indonesia: An Alternative History* by Malcolm Caldwell and Ernst Utrecht.
American Geographical Society, *China: Land of Famine* by J A Harrison.
AMS Press, *Indian Records Series: Bengal in 1756–1757* by C S Hill.
Associated Publishers Inc, *The History of the Burgis* by Mariel Francisco and Fe Maria Arriola; *The Philippine Islands 1493–1898* by Emma Helen Blair and James Alexander Robertson.
Basil Blackwell, *China and the West* by Wolfgang Franke.
Geoffrey Bles, *A Tale from Bali* by Vicki Baum.
The Bodley Head, *Nehru: The First Sixty Years* by Dorothy Norman; *Jawaharlal Nehru: an autobiography.*
N V Boekhundel & Drukkerij voorheen E J Brill, Leiden, *The Social Condition of the British Community in Bengal 1757–1800* by Suresh Chandra Ghosh.
The British Library, extract from *The Graphic.*
Cambridge Book Co, *Patterns of Civilization: Asia* by Peter G Kontos.
Cambridge University Press, *The Cambridge Economic History of India* edited by Dharma Kumar; *The Pattern of Imperialism: The United States, Great Britain, and the late-industrializing world since 1815* by Tony Smith; *The Christian Century in Japan 1549–1650* by C R Boxer; *The Correspondence of G. E. Morrison, 1895–1912* by Lo Hui-Minh; *The Emergence of Indian Nationalism* by Anil Seal; extract from 'The Seavoyage Controversy and the Kayasthas of North India, 1901–1909' by Lucy Carroll in *Modern Asian Studies.*
Frank Cass & Co Ltd, *Ceylon Under British Rule 1795–1932* by Lennox A Mills; *The British Impact on India* by Sir Percival Griffiths.
Cassell & Co, *The India We Served* by Sir Walter Roper Lawrence.
Chatto & Windus and The Hogarth Press, *Apprentice to Power: India 1904–1908* by Sir Malcolm Darling; *Memoirs of a Bengal Civilian* by John Beames.
Chugh Publications, *Political Trouble in India* by H W Hale; *Among Indian Rajahs and Ryots* by Sir Andrew H L Fraser.
Collier Associates, *The Spanish War: An American Epic—1898* by G J A O'Toole.
Columbia University Press, *Sources of Chinese Tradition* edited by Wm Theodore de Bary *et al*; *Sources of Indian Tradition* by Wm Theodore de Bary; *Sources of Japanese Tradition* by Wm Theodore de Bary *et al.*
Constable & Co Ltd, *India as I Knew it* by Sir Michael O'Dwyer.
Cornell University, *Rajah's Servant* by A B Ward; *The Communist Uprisings of 1926–1927 in Indonesia: Key Documents* by Harry Benda and Ruth T McVey.
Cornell University Press, *Mongkut, the King of Siam* by Abbot Low Moffat.
Deep Publications, *Disorders Inquiry Committee 1919–1920 Report* by Lord Hunter, Chairman.
J M Dent & Sons, *Portuguese Voyages 1498–1663* edited by Charles David Ley; *Lord Jim* by Joseph Conrad; *The Discoverers* by Daniel Boorstin.
Doubleday & Co Inc, *Decolonization: The Administration and Future of the Colonies, 1919–1960* by Rudolf von Albertini; *Greville Diary* edited by Philip Whitwell Wilson; *Karl Marx on Colonialism and Modernization* by Shlomo Avineri.
Eastern Universities Press Sdn Bhd, *Memories of a Nonya* by Queeny Chang.

ber & Faber Ltd, *Raffles* by Maurice Collis.
arvin E Gettleman, (ed), *Vietnam.*
arcourt Brace Jovanovich Inc, *A Matter of Honour* by Philip Mason.
arvard University Press, *China's Response to the West: A Documentary Survey 1839–1923* by Ssu-Yu Teng and John K Fairbank; *A Source Book in Geography* by George Kish.
M Heath, Estate of the late Sonia Brownell Orwell and Martin Secker & Warburg Ltd, *Burmese Days* by George Orwell.
C Heath & Co, *The End of the European Empire: Decolonization after World War II* edited by Tony Smith.
illiam Heinemann Ltd, *Collected Short Stories* by Somerset Maugham.
er Majesty's Stationery Office, *The Transfer of Power 1942–7* edited by Nicholas Mansergh.
odder & Stoughton Ltd, *The Memoirs of Cordell Hull.*
e Hokuseido Press, *The Autobiography of Fukuzawa Yukichi* transl by Eiichi Kiyooka.
uman Resource Development Press, *Life and Ritual in Old Siam: Three Case Studies of Thai Life and Customs* by Phya Anuman Rajadhon.
ico Publishing House, *The Autobiography of an Unknown Indian* by Nirad Chaudhuri.
ichael Joseph, *Bhowani Junction* by John Masters.
e Macmillan Co, *French Indo-China* by Virginia Thompson.
acmillan Publishers Ltd, *Disparities in Economic Development since the Industrial Revolution* edited by Paul Bairoch and Maurice Lévy-Leboyer; *Sociology of 'Developing Societies'* edited by Hamza Alavi and John Harriss; *The Political Economy of the Raj 1914–1947* by B R Tomlinson; *Asia—the Winning of Independence* by Robin Jeffrey; *Curzon: A Most Superior Person: A Biography* by Kenneth Rose; *The Imperial Frontier in the Tropics, 1865–75* by W David McIntyre.
ichigan State University, *Sir Saiyyid Ahmad Khan's History of the Bijnor Rebellion* transl by Hafeez Malik.
ntre for South-East Asian Studies, Monash University, *Indonesian Nationalism and Revolution: Six First-hand Accounts* edited by H Feith.
e *New Republic*, extract from 'Echoes of a Revolt' by Raymond Postgate.
eana Publications Inc, *China and the Foreign Powers: The Impact of and Reaction to Unequal Treaties* by William L Tung.
iental Publishers, *Political Trouble in India: 1907–1917* by James Campbell Ker.
xford University Press, *A Nation in Making* by Sir Surendranath Banerjea; *World Christian Encyclopedia* edited by David Barrett; *The Making of Modern South-East Asia* by D J M Tate; *The Making of British India 1756–1858* by Ramsay Muir; *Honourable Intentions: Talks on the British Empire in South-East Asia* edited by Paul H Kratoska; *Wavell: The Viceroy's Journal* edited by Penderel Moon; *The Moral and Political Writings of Mahatma Gandhi* edited by Raghavan Iyer; *The Nabobs* by Percival Spear; *The Evolution of India and Pakistan 1858–1947* by C H Phillips.
Paragon Book Gallery Ltd, *A Cycle of Cathay* by W A P Martin.
Penguin Books Australia Ltd, *This Earth of Mankind* by Pramoedya Toer.
The Penguin Group, *The Economics of Imperialism* by Michael Barratt Brown; *The Dragon Wakes; China and the West, 1793–1911* by Christopher Hibbert.
Frederick A Praeger Publishers, *Vietnam: a Dragon Embattled* by Joseph Buttinger.
Prentice-Hall Inc, *The Emergence of Modern Southeast Asia: 1511–1957* edited by John Bastin; *The Political Awakening in India* edited by John McLane.
Her Majesty's Stationery Office, Public Record Office (UK), extract from CAB 66/26.
Routledge, Chapman & Hall Ltd, *Travel and Travellers of the Middle Ages* by Arthur Percival Newton.
Sage Publications India Pvt Ltd, extract from 'Malaria and Mortality in Bengal, 1840–1921' by Ira Klein in *Indian Economic and Social History Review.*
Schenkman Publishing Co, *The Westernization of Asia: A Comparative Political Analysis* by Frank C Darling.
Martin Secker & Warburg Ltd, *My Diaries: Being a Personal Narrative of Events 1888–1914* by Wilfred Scawen Blunt; *Daddyji* by Ved Mehta.
Colin Simmons, 'Labour and Capital in the Gold Mining Industry in India'.
Singapore University Press Pte Ltd, extract from 'Singapore: Its Growth as an Entrepot Port, 1819–1914' by Lin Ken Wong in *Journal of South East Asian Studies.*
Stanford University Press, *The Japanese Discovery of Europe, 1720–1830* by Donald Keene; *The May Fourth Movement* by Chow Tse-Tsung.
University of California Press, *Tilak and Gokhale: Revolution and Reform in the Making of Modern India* by Stanley A Wolpert; *Agrarian Conditions in Northern India* by Elizabeth Whitcombe; *The Development of Capitalism in Colonial Indochina (1870–1940)* by Martin J Murray.
University of Chicago Press, extract from 'The Tools of Imperialism' by D R Headrick in *Journal of Modern History.*
East-West Centre Press, University Press of Hawaii, *The Thai Bureaucracy: Institutional Change and Development* by William J Siffin.
University of Malaya Press, *British Intervention in Malaya 1867–1877* by C Northcote Parkinson.

University of Minnesota Press, *Rival Empires of Trade in the Orient 1600–1800* by Holden Furber.
University of Queensland Press, *Indonesia: Selected Documents on Colonialism and Nationalism, 1830–1942* edited by C L M Penders.
University of Pennsylvania Press, *Agricultural Trends in India, 1891–1947: Output, Availability, and Productivity* by George Blyn.
Unwin Hyman Ltd, *Problems of Empire: Britain and India 1757–1813* by P J Marshall; *Indian Dust: Being Letters from the Punjab* by P E Richards; *The Dutch Seaborne Empire 1600–1800* by C R Boxer.
Vikas Publishing House Pvt Ltd, *Crippled Minds: An Exploration into Colonial Culture* by Susantha Goonatilake; *From Sepoy to Subedar: being the Life and Adventures of Subedar Sita Ram* edited by James Lunt.
Walker & Co, *Imperialism* by Philip D Curtin.
George Weidenfeld & Nicolson, *The Pelican Economic History of Britain*; by E J Hobsbawm.
C & J Wolfers Ltd, *Japan Versus Europe: A History of Misunderstanding* by Endymion Wilkinson.
Yale University Press, *Patterns of Vietnamese Response to Foreign Intervention: 1858–1900* by Truong Buu Lam.

We also thank the following people and organisations for permission to reproduce and/or supplying photographs and illustrations:

Cover illustrations
Front: Kluwer Academic Publishers (Dutch ships leaving Java); Popperfoto (peddling Christian literature in China); National Maritime Museum Greenwich (Vasco da Gama); The Hulton Picture Company (An Afternoon in the Plains); Uitgerverij T Wever B V (An English factor being tortured); The Hulton Picture Company (His Highness the Maharaja); The British Library (Royal seals of Siam).
Back: Popperfoto (Rickshaws in India); Popperfoto (Bentinck); Popperfoto (Summer Palace).

Text illustrations
Department of Geography, Monash University (pp 1, 7, 149, 183, 185, 191).
India Office Library and Records (p 20).
International Society for Educational Development (pp 103, 110, 116).
Kedleston Hall (p 84).
Kluwer Academic Publishers, Dordrecht, Holland (p 3).
Mary Evans Picture Library (p 21).
National Maritime Museum Greenwich (p 9).
Nehru Memorial Museum and Library (p 123).
Popperfoto (pp 13, 44, 53, 58, 63, 65, 68, 76, 79, 126, 147, 161, 162, 194, 195 (top)).
Punch Publications Ltd (pp 155, 169).
Studio Editions Ltd, London (pp 34, 48).
Thames & Hudson Ltd (p 28).
The British Library (pp 90, 113–14, 140).
The Granger Collection, New York (p 41).
The Hulton Picture Company (pp 18, 36, 55, 93, 180).
The *Illustrated London News* (pp 33, 73).
The Library of Congress (p 136).
Topham Picture Source (pp 168, 177).
Uitgerverij T Wever B V (p 14).
Victoria and Albert Museum (p 188).

Disclaimer
Every effort has been made to trace the original source material contained in this book. Where the attempt has been unsuccessful, the publishers would be pleased to hear from copyright holders to rectify any omissions.